The Oil & Gas Industry
A Nontechnical Guide

THE
OIL & GAS
INDUSTRY A NONTECHNICAL GUIDE

JOSEPH F. HILYARD

Copyright© 2012 by
PennWell Corporation
1421 South Sheridan Road
Tulsa, Oklahoma 74112-6600 USA

800.752.9764
+1.918.831.9421
sales@penwell.com
www.pennwellbooks.com
www.pennwell.com

Director: Mary McGee
Managing Editor: Stephen Hill
Production Manager: Sheila Brock
Production Editor: Tony Quinn
Book Designer: Susan E. Ormston
Cover Designer: Charles Thomas

Library of Congress Cataloging-in-Publication Data

Hilyard, Joseph.
  The oil and gas industry : a nontechnical guide / Joseph F. Hilyard.
      p. cm.
  Includes bibliographical references and index.
  ISBN 978-1-59370-254-0
1. Petroleum industry and trade. 2. Petroleum engineering. I. Title.
  HD9560.5.H5125 2012
  338.2'7282--dc23

                                                    2012007391

Printed in the United States of America

5 6 7 8 9   21 20 19 18 17

To my two sons, Jonathan and Aaron,
whom I hereby direct to use wisely their share
of the petroleum products and natural gas
brought to market in the 21st century.

Contents

Preface

This book provides a nontechnical overview of what is commonly referred to as the petroleum industry, focusing on oil and natural gas, as well as their derivative products. My goal has been to create an engaging and accessible introduction to this critically important global industry to help those without a technical background who are either new to the industry or simply interested in how it operates.

This book guides the reader through the series of typical decisions made, actions taken, and equipment and processes used to bring petroleum products and natural gas to world markets. I have tried to present information about these various activities in broad but also clear and accurate terms. However, I have not attempted to describe every operational facet or technical aspect; this level of detail is available from a variety of other sources, a number of which are listed in appendix B.

The word "petroleum" is based on the Latin *petrus* (rock) and *oleum* (oil). Petroleum can range in color from nearly colorless to jet-black. It can be thinner than water or thicker than molasses, and its density can vary from that of a light gas to that of a heavy asphalt. For the purposes of this book, the term *petroleum* is used to collectively describe oil and natural gas, whereas the *petroleum industry* includes the various entities that perform the range of activities noted in the bulleted list below, as well as those that support those activities.

Chapter 1 describes the geologic processes and structures related to the formation of both crude oil and natural gas within the earth, as well as its movement (migration) that results in the creation of commercially exploitable accumulations. Chapters 2 and 3 then focus individually on oil and gas, respectively, providing basic information about their composition, the location of major oil and gas resources around the world, the range of products created from oil and gas, and current patterns of, as well as future projections for, production and use.

Chapters 4 through 12 address petroleum industry operations:

- Searching for and evaluating petroleum resources
- Drilling and completing wells to tap promising formations (onshore and offshore)
- Managing oil and gas production
- Transporting oil
- Transporting natural gas in both gaseous form (by pipeline) and in liquefied form
- Converting oil and gas into various products

Chapters 13 through 15 address several business-related issues: the structure of the industry, the dynamics of petroleum trading, and the challenges facing the industry. Finally, three appendices provide a listing of terms, abbreviations, and acronyms used in the book; suggested further reading; and organizations that can provide further information on many of the topics covered in this book.

I hope this book helps the reader understand the petroleum industry and makes clear the ingenuity and skill of its millions of professionals worldwide.

1 Origins of Oil and Gas

A discussion of the operations of the petroleum industry—including production and the extensive slate of useful products—must begin with a review of the origins of the raw material comprising the foundation of the industry's value chain. In the case of petroleum, the origins of crude oil and natural gas can be traced back millions—in fact, tens of millions—of years ago, to the seabed of ancient oceans.

A Brief Overview

For hydrocarbons to accumulate, three conditions must be met. First, a sedimentary basin must be created—the result of movement of the earth's crust, which creates large depressions into which sediments from surrounding elevated areas are transported over time.

Second, the sediments laid down in such basins must contain a high level of organic material. This organic-rich matter becomes part of the sedimentary material to create what is called *source rock*.

Third, over millions of years, the effects of elevated temperature and pressure must be sufficient to convert the material in the source rock into oil and gas. *Maturity* describes the degree to which petroleum generation has occurred. Heavy, thick oil is considered immature, having been generated at relatively low temperature. Mature oil—lighter or less viscous—forms at higher temperature.

In an important subsequent process called *migration*, the hydrocarbons must move out of the source rock through cracks, faults, and fissures and into porous and permeable *reservoir rock*. Finally, that reservoir rock must be configured (as a result of prior geologic activity) in a way that immobilizes the hydrocarbons within structures called *traps*, allowing oil and gas to accumulate in sufficient volumes to warrant commercial exploitation.

Subsea Burial

To begin a more detailed look at the pathway to petroleum, it can be said that petroleum geochemists and geologists widely agree that crude oil is derived from ancient organic matter—ranging from single-celled plankton to more-complex aquatic plants (e.g., algae) and even invertebrates and fish—laid down and then buried and preserved in sediments eons ago at the bottom of ancient oceans. Organic material carried into the oceans by rivers also is deposited and buried in the same manner.

The buried organic matter undergoes a transformation over millions of years. Initial microbial action (in the presence of oxygen dissolved in seawater) returns part of the sediment's carbon to the atmosphere as carbon dioxide. However, as the sediment layer gets thicker, subsequent bacterial processes at work in deep seabed mud (with little or no oxygen present in the mud itself or the water immediately above it) convert the remaining organic matter into a waxy material called *kerogen*.

Kerogen is a complex mixture of large organic molecules whose appearance and characteristics depend on the kind and concentration of materials—such as algae, plankton, bacteria, pollen, resin, and cellulose—of which it is composed. It is from kerogen that oil and gas are generated.

Organic matter was not laid down evenly through the various geologic eras (table 1–1). Accumulations were clearly concentrated in a limited number of intervals whose duration was determined by movements in the earth's crust and by climatic changes. Different periods contributed the following approximate shares of the world's kerogens:

- Middle Cretaceous (about 100 million years ago): nearly 30%
- Late Jurassic (150 million years ago): 25%
- Late Devonian (350 million years ago): less than 10%
- Silurian (420 million years ago):less than 10%
- Early Cambrian (550 million years ago): less than 10%

Table 1–1. Geologic timescale

Period	Epoch	Start of period, millions of years ago
Cenozoic era:		
Quaternary	Holocene	0.01
Quaternary	Pleistocene	1.6
Tertiary: Neogene	Pliocene	5.3
Tertiary: Neogene	Miocene	23.7
Tertiary: Paleogene	Oligocene	36.6
Tertiary: Paleogene	Eocene	57.8
Tertiary: Paleogene	Paleocene	66.4
Mesozoic era:		
Cretaceous		144
Jurassic		208
Triassic		245
Paleozoic era:		
Permian		286
Carboniferous: Pennsylvanian		320
Carboniferous: Mississippian		360
Devonian		408
Silurian		438
Ordovician		505
Cambrian		570
Proterozoic era		2,500
Archean era		3,800
Hadean era		4,550

Source: U.S. Geological Survey.

The total amount of organic matter stored in the earth's crust has been estimated at 10 quadrillion (10^{16}) tons, of which approximately 1% (100 trillion [10^{14}] tons) is in organic-rich rocks—principally, shales containing at least 3% organic matter.

Hydrocarbon Generation within Source Rock

Kerogen concentrations as low as 1%–3% are generally sufficient to make source rock (typically shales and limestones) suitable for commercial exploitation of crude oil and natural gas. Black shale is the most common kind of source rock. Oil source rocks can contain up to 40% organic matter, and the level approaches 100% for some types of coal.

Temperature plays a key role in the generation of oil and gas from kerogen. As the organic-rich source rock undergoes progressive burial (i.e., as additional sediments are laid down above it), the rock becomes hotter. This phenomenon reflects what is called the *geothermal gradient* of the earth. From the surface to a depth of about 60 meters (200 feet), ground temperature a relatively constant 11°C (55°F). From that point down to about 122 meters (400 feet), the gradient is variable, owing to atmospheric influences and circulating groundwater.

Below 400 feet, temperature rises steadily with depth, though the rate of increase varies with location. For tectonically stable shield areas and sedimentary basins, a typical figure is 1.5–2°C per 30.5 meters (1.5–2°F per 100 feet).

The term *oil window* is used to describe the range of temperature or depth within which most of oil's complex constituents are produced (fig. 1–1). This window is typically 80–220°C (176–428°F) or 2,200–5,500 meters (7,200–18,000 feet).

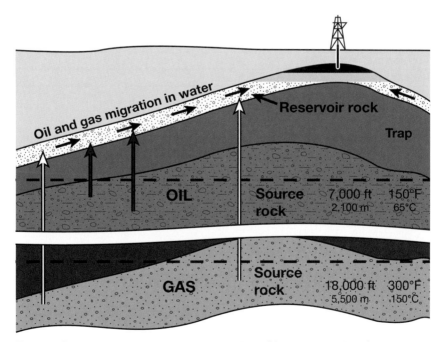

Fig. 1–1. Generation and movement of oil and gas (*Source:* Hyne, N. J.)

Other factors that can affect the rate of oil generation are pressure (imposed by overlying rock and sediment), the presence of heat-tolerant bacteria that act on the oil, and the presence of hydrogen and oxygen (from water and surrounding minerals). As these various drivers exert their influence, the kerogen in source rock undergoes conversion to petroleum in a process called *maturation*. Below a temperature of about 50°C (122°F), small amounts of kerogen begin to be transformed into oil. Peak conversion occurs at about 100°C (212°F).

If the temperature rises above 130°C (266°F) for even a brief time, then crude oil itself begins to break into smaller molecules, and gas starts to be produced. At first, this will be *wet* gas and condensate, with high levels of relatively heavy hydrocarbons. With further increases in temperature, the gas will become more *dry*, containing more of the lighter hydrocarbon gases.

Abiogenic hydrocarbons

About 60 years ago, Russian and Ukrainian scientists postulated an alternative mechanism for the formation of hydrocarbons. The *abiogenic* (inorganic) process they envisioned does not involve biological material, but rather the transformation of hydrogen and carbon into oil and gas deep within the outer mantle of the earth. This is postulated to take place at temperatures greater than 1,500°C (2,730°F) and pressures 50,000 times that at the earth's surface— conditions found at a depth of about 65 miles.

Austrian-born Thomas Gold (1920–2004), a leading U.S. astronomer and geoscientist, was perhaps the best-known Western advocate for the theory.

Despite tantalizing results from limited laboratory and field experiments, it remains highly speculative.

Migration to Reservoir Rock

Oil and gas typically move around within (and are eventually expelled from) source rock after generation, typically driven by the pressure of overlying rock layers and aided by the presence of faults and cracks in the source rock and nearby rock layers. In fact, virtually all commercially viable oil reservoirs result from migration that takes the hydrocarbons away from the source rock and into *reservoir rock*.

Oil very rarely collects in large underground pools of liquid, accumulating instead in the pores of highly permeable reservoir rock. Permeable rock has extensive and well-connected pores that enable (eventual) substantial hydrocarbon flow to a drilled wellbore.

Reservoir rock with good porosity and permeability is generally classified as either a *clastic* or a *carbonate* system. Clastic sediments are formed from fragments of various rocks that were transported and redeposited to create new formations. Sandstones, siltstones, and shales are the most common types. Notable clastic depositions are found in river delta regions, such as along the U.S. Gulf Coast, several Venezuelan coastal fields, the Niger River delta in Africa, and the south Caspian Sea.

Carbonate rock, in contrast, is typically formed by a chemical reaction between calcium and carbonate ions in shallow seas, or by a process called *biomineralization*. The most notable example of the latter is the creation of large reefs by marine coral. Carbonate rock also can build up from the constant rain of tiny shell fragments that falls to the sea floor from microorganisms living in the water above. Limestone and dolostone are typical carbonates.

Hydrocarbon Traps

In a promising hydrocarbon-bearing formation, oil and gas have migrated into what is called a *trap*. There are several types, all created by prior deformation of the earth's crust. Highly impermeable rocks above and around the trap seal it in a way that prevents any further significant movement of hydrocarbons upward or laterally. There are three basic types of hydrocarbon traps: structural, stratigraphic, and combination (fig. 1–2).

Structural traps

Structural traps are formed by tectonic processes—movement of the rock plates that comprise the top of the earth's crust—that deform underlying rock layers. One common type is the anticline, a smooth, archlike fold. By some estimates, nearly 80% of the world's largest oil reservoirs are found in anticlinal traps.

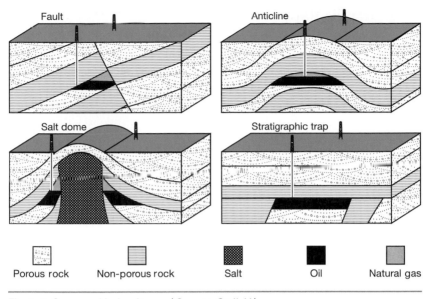

Fig. 1–2. Common kinds of traps (*Source:* Smil, V.)

A second type of structural trap is the fault trap, created by the displacement of rock layers (strata) relative to each other. When rock strata have moved mostly horizontally, a *strike-slip* fault results (e.g., the San Andreas Fault, in California). Movement that is mostly vertical and downward creates a normal fault, while vertical upward movement results in a thrust (or reverse) fault. Hydrocarbon accumulations are typically associated with normal and thrust faults. These basic fault types are shown in fig. 1–3.

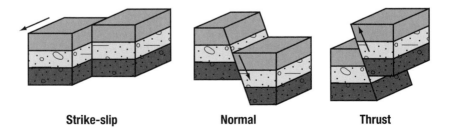

Fig. 1–3. Fault types (*Source:* Adapted from U.S. Geological Survey)

Stratigraphic traps

Stratigraphic traps are created when a seal or barrier is formed above and around an oil- or gas-bearing formation by sedimentary deposition of impermeable rock. Primary stratigraphic traps include channels and barriers of sandstone in a river delta area, carbonate slopes, coral reefs, and clay-filled channels of dolomite or calcite.

Major oil fields with stratigraphic traps include Prudhoe Bay (Alaska), East Texas, and a supergiant field along the east coast of Lake Maracaibo, in Venezuela.

Combination traps

Combination traps are formed by a combination of processes that occurred in the sediments during the time of deposition of the reservoir bed. They are also formed by tectonic activity that occurred in the reservoir beds after their deposition.

One example of a combination trap is associated with a salt dome—a mass of lighter salt that has pushed upward through heavier surrounding rock and sediments. Salt beds were formed by the natural evaporation of seawater from an ancient enclosed basin, and the resultant salt layer was then buried by successive layers of sediments over geologic time.

The upward push creates an anticlinal type of folding, with reservoir rock frequently found draping the flanks of the salt dome to create conditions for hydrocarbon trapping. Salt itself is impermeable to oil and gas and can contribute to trap creation. As described in later chapters, a cavity can be created in the salt formation itself, comprising an effective storage cavern for hydrocarbon products.

Three conclusions can be drawn from this discussion of hydrocarbon generation, migration, and trapping. First, by some estimates, an average of only about 10% of all the gas and oil that forms in a sedimentary basin ever reaches a trap. The remainder never moved from the source rock, is lost during migration, or seeps from the earth's surface.

Second, oil may in some cases be found above gas. Even though figures 1–2 and 1–3 show gas accumulating above oil in various traps, the situation can be more complicated.

Third, as will be discussed in more detail later (see chap. 7, on hydrocarbon production), water is virtually always found in the pore spaces of reservoir rock, intermingled with oil and gas. For this reason, most wells pump not only oil and gas but also mineral-laden water called *brine*.

2 Oil Overview

Petroleum in its most common liquid form is referred to as *crude oil*. It is extracted from the earth by drilling into geologic structures (described in chap. 1) whose properties are conducive to the aggregation of oil.

General Composition of Crude Oil

Petroleum consists mostly of hydrocarbon molecules, themselves made up of various combinations of hydrogen and carbon atoms. The simplest hydrocarbon molecule—one carbon atom bonded to four hydrogen atoms (CH_4)—is called *methane* and is the primary component of natural gas. More details about this form of hydrocarbon are given in chapter 3.

Crude oil is a mixture of a very large number of hydrocarbons with arcane names such as *alkanes* (or paraffins), *cycloalkanes* (or naphthenes), *aromatic* hydrocarbons, and *asphaltenes*. A major distinguishing characteristic is the number of carbon atoms they contain. For example, asphalt contains 35 or more carbon atoms per molecule, while pentane (used to make gasoline) has just 5. (Both are classified as alkanes.) The type, variety, and structure of the hydrocarbon molecules in crude oil determine its physical and chemical properties, such as color, thickness (viscosity), and boiling point. Crude oil also may contain nitrogen, oxygen and sulfur, plus trace amounts of metals such as iron, nickel, copper, and vanadium. In general terms, crude oil contains about 83%–87% carbon, 10%–14% hydrogen, 0.1%–2% nitrogen, 0.1%–1.5% oxygen, and less than 0.1% metals.

Conventional Crude Oil

Crude oil is often found along with natural gas and saline (salty) water. The lighter natural gas forms a gas cap over the petroleum, while the saline water is heavier than most forms of crude and generally sinks beneath it. Conventional oil resources are those whose location and extent have been quite well established, that can be produced through wellbores using relatively standard equipment and procedures, and that typically require minimal processing prior to sale.

Classification scheme

The petroleum industry uses three major parameters to classify crude oil:

- Geographic location in which it is produced (which affects the cost of transporting the crude to a refinery).

- API gravity (an oil industry measure of density; API is the American Petroleum Institute). Light crude oil has relatively low density; heavy crude has high density. Oil with an API gravity (expressed as °API) below 10.0 is classified as extraheavy.

- Sulfur content. Crude is generally called *sweet* if it contains relatively little sulfur or *sour* if it contains substantial amounts.

Light crude oil is more desirable than heavy oil because it produces a higher yield of gasoline, a highly valued petroleum product for transportation use. Sweet oil commands a higher price than sour oil because it has fewer environmental problems and requires less refining to meet sulfur-content standards imposed by buyers. Table 2–1 shows the API gravity and sulfur content of some benchmark crude oil (described in the next section).

Each crude oil has unique molecular characteristics that are evaluated by a process called *assay analysis*, carried out in a laboratory.

Benchmark crudes

Oil from an area in which its molecular characteristics have been determined is used as a pricing reference, or *benchmark*, in global oil markets. Some common reference crudes are:

- *West Texas Intermediate* (WTI), a very high quality sweet, light oil delivered at Cushing, Oklahoma, for North American oil. Cushing is the delivery point for WTI traded on the New York

Table 2–1. Properties of selected crude oils

Crude type	Country	Gravity, °API	Sulfur content,%
Arabian Light	Saudi Arabia	33.4	1.80
Bachequero	Venezuela	16.8	2.40
Bonny Light	Nigeria	37.6	0.13
Brass River	Nigeria	43.0	0.08
Dubai	Dubai (UAE)	32.5	1.68
Ekofisk	Norway	35.8	0.18
Iranian Light	Iran	33.5	1.40
Kuwait	Kuwait	31.2	2.50
North Slope (Alaska)	US	26.8	1.04
WTI[a]	US	38–40	0.24

Source: Hyne, N. J.
[a] West Texas Intermediate.

Mercantile Exchange (NYMEX), the most widely traded oil futures contracts in the world.

- *Brent Blend*, made up of 15 oils from fields in the Brent and Ninian systems in the East Shetland Basin of the North Sea. Oil production from Europe and Africa, as well as Middle Eastern oil flowing to the West, tends to be priced using this benchmark.

- *Dubai-Oman*, used as benchmark for Middle East sour crude flowing to the Asia–Pacific region.

- *OPEC Reference Basket*, a weighted average of oils and blends from the 12 nations that make up the Organization of the Petroleum Exporting Countries.

- *Midway-Sunset Heavy*, by which heavy oil in California is priced. Midway-Sunset is a large oil field in Kern County, California.

Declining amounts of these benchmark oils are being produced each year, so other oils are more commonly what is actually delivered. For example, while a reference price may be for WTI delivered at Cushing, the actual oil being traded may be a Canadian heavy oil delivered at Hardisty, Alberta.

In early February 2011, the *Financial Times* noted the continuing decline in the dominance of the WTI benchmark in global oil pricing. Because of increasing crude inventories at Cushing, the WTI price had weakened substantially compared to crudes of comparable quality, such as Brent Blend and Light Louisiana Sweet.

At one point in early February 2011, Brent's premium over WTI reached a record level of more than $16 per barrel. Such discrepancies upset the plans of energy investors and traders. In fact, in 2009, Saudi Arabia dropped WTI as its benchmark for pricing oil for sale to customers in the United States.

Crude oil prices

A brief word is warranted about crude prices, which jumped about 30% between January and early April 2011. (On April 11, 2011, the Brent price reached a 2.5-year high of $127.02 per barrel, with WTI at $112.79.) Driving fears of further increases were concerns about chaotic geopolitics in several oil-producing countries, including:

- Actions aimed at ousting long-standing regimes in North Africa—from Tunisia and Egypt to Libya, Syria, and Algeria— as well as unrest in Yemen and Saudi Arabia.

- Continuing civil disorder in Nigeria, a major global source of highly prized light, sweet crude oil.

- Increased domestic spending in some of these nations (e.g., $125 billion in Saudi Arabia) in an attempt to quell discontent, which might then be made up by charging more for exported oil.

- Continued use of domestic fuel subsidies in many oil-producing countries, which tend to drive up internal demand and reduce oil volumes available for export.

Also contributing to concerns about future price and supply was rising oil demand in China, already the world's second-largest oil consumer in 2010. Some analysts were projecting that oil consumption in China would top 10 million barrels per day (b/d) in 2011.

A counterbalance to upward price pressure was a slowing global economy. By mid-August 2011, the crude oil futures contract price on the NYMEX had fallen back to $80–$85 per barrel.

However, Middle Eastern geopolitics again pushed up oil prices in early 2012. On January 23, the price of NYMEX crude for March delivery rose to $99.58 per barrel, driven in part by a decision that day by European Union governments to ban imports of oil from Iran (effective July 1, 2012) and to freeze Europe-based assets of Iran's central bank. (Approximately 20% of Iranian oil exports go to EU nations.)

The EU action was part of an international campaign to pressure Iran to resume talks on the country's nuclear development program.

In response, some Iranian lawmakers threatened to close the Strait of Hormuz, a narrow passage between Iran and Oman and a critical route for tankers that carry oil from the Persian Gulf to world markets.

Unconventional Crude Oil

Several types of oil resources are called unconventional, to distinguish them from oil that can be extracted using traditional oil field methods. These include *tar sands* (also called *oil sands*) and *shale oil*.

Tar sands/oil sands

Crude oil is sometimes found in semisolid form, mixed with sand and water. Major deposits of oil of this type are found in the Athabasca region of northeastern Alberta. Tar sands contain *bitumen*—a kind of heavy crude oil. The sticky, black, tarlike material is so thick that it must be heated or chemically diluted before it will flow. Oil-eating bacteria have destroyed some of the lighter fractions of crude oil in such oil sands, leaving behind the heavier bitumen fractions.

Venezuela also has large amounts of crude trapped in sands in that nation's Orinoco region. These hydrocarbons have a somewhat lower viscosity (i.e., are less thick) than those in Canada and are usually called *heavy* or *extraheavy* oil.

Combined, Canada and Venezuela contain an estimated 3.6 trillion barrels (570 billion cubic meters) of bitumen and extraheavy oil. This represents about twice the volume of the world's reserves of conventional oil.[1]

Shale oil

Another unconventional resource, shale oil is found in shale source rock that has not been exposed to heat or pressure long enough to convert trapped hydrocarbons into crude oil. According to the U.S. Department of Energy (DOE), oil shales are not technically shales and do not really contain oil. Rather, they are usually relatively hard rocks called *marls*—composed primarily of clay and calcium carbonate—containing the waxy substance called kerogen (described in chap. 1). The trapped kerogen can be converted into crude oil using heat and pressure to simulate natural processes.

Oil shales are found in many countries, but the United States has the world's largest deposits. These are located chiefly in the Green River formation, which covers portions of Colorado, Utah, and Wyoming.[2] DOE has estimated that as much as 800 billion barrels of oil are recoverable from that formation.

In addition, from 2008 to 2011, small and midsize companies stepped up exploration and production in several other regions, including the Bakken oil-bearing shale along the North Dakota/Montana border, the Permian Basin in West Texas, and the Eagle Ford formation in South Texas. The companies used two main techniques previously developed to extract natural gas from shale formations: the drilling of long horizontal wells to reach additional oil deposits; and hydraulic fracturing (fracing), in which water mixed with chemicals and other materials is pumped at high pressure into the ground to crack the rock and make it easier for oil to flow. Analysts have estimated that the United States could be producing an additional 2.5 million b/d of oil by 2016, with the increase coming equally from deepwater fields in the Gulf of Mexico and from new onshore sources.

As the world's reserves of conventional light and medium oil are depleted, oil refineries are investing in the more-complex and expensive systems needed to process increased volumes of heavy oil and bitumen. Heavier crude oils have too much carbon and not enough hydrogen, so these systems generally involve removing carbon or adding hydrogen, to convert the longer, more-complex molecules in the oil to the shorter, simpler ones that characterize end-product fuels. (More details of these processes are presented in chap. 11.)

Resources and Reserves

A brief discussion of oil resources and reserves is warranted here. *Resources* are hydrocarbons that may or may not be produced in the future. An initial estimate of the magnitude of the resource in a yet-to-be-drilled prospect may be made. However, appraisal (by drilling delineation wells or acquiring more seismic data, as described in chap. 4) is needed to confirm the size of a field and pave the way for development and production. Once the relevant government body gives an oil company a production license for field development, the reserves in the field can be formally booked.

According to the Society of Petroleum Engineers (SPE), reserves are the quantities of crude oil estimated to be commercially recoverable by application of development efforts to known (discovered) accumulations, from a given date forward, under defined conditions. The total estimated amount of oil in a reservoir, including both producible and nonproducible oil, is called *oil in place* (OIP). However, because of reservoir characteristics and the limits of extraction technologies, only a fraction of this oil can be brought to the surface. Only this producible fraction comprises the reserves. The ratio of producible oil reserves to total OIP for a given field is often referred to as the *recovery factor,*

Reserves are further categorized by the level of certainty associated with the estimates of their magnitude. The following are the three categories in widest use:

- *Proved* reserves. These are considered reasonably certain to be producible using current technology at current prices, with current commercial terms and government consent. This category is also sometimes called 1P, but others in the industry refer to it as *P90*—that is, having a 90% certainty of being produced.

- *Probable* reserves. These are considered to have a reasonably probable chance of being produced using current or likely technology at current prices, with current commercial terms and government consent. This category is sometimes called *P50*—that is, having a 50% certainty of being produced—and is also known in the industry as *2P*, meaning "proved plus probable."

- *Possible* reserves. These have a chance of being developed under favorable circumstances. Some industry specialists refer to this as *P10*—that is, having a 10% certainty of being produced—while others use the term *3P*, for "proved plus probable plus possible."

Experience has shown that initial estimates of the size of newly discovered oil fields are usually too low. The term *reserve growth* refers to the increases in estimated ultimate recovery that occur as oil fields are developed and produced.

Known global oil reserves are estimated at about 1.2 trillion barrels (without oil sands) by the Energy Information Administration (EIA) of the DOE. The agency raises that figure to 3.74 trillion barrels if oil sands are included.

Barrels of oil equivalent (boe)

Energy companies and others sometimes use the parameter *barrels of oil equivalent* (boe) as a way to report oil plus gas reserves or production as a single figure. Oil and gas from various sources release different amounts of energy when burned, but SPE provides the following information, based on typical heating values.

Burning one barrel (42 gallons) of oil releases 5.8 million British thermal units (Btu). One Btu is the energy needed to raise the temperature of one pound of water one degree Fahrenheit at one atmosphere of pressure.

A typical figure for the heating value of natural gas is 1,025 Btu per cubic foot, according to SPE. Therefore, burning 5,658.53 cubic feet of gas would yield an equivalent amount of energy to burning one barrel of oil (5.8 million divided by 1,025). A company with gas reserves of 5,658.53 cubic feet and oil reserves of 10 barrels could then report its total reserves as 11 boe.

The U.S. Geological Survey defines 1 boe as roughly equivalent to 6,000 cubic feet of typical natural gas. The French oil company Total uses a figure of 5,487 cubic feet.

Table 2–2 lists those nations with the largest estimated oil reserves as of early 2012, as well as data on number of oil wells and oil production.

Figure 2–1 provides another perspective on global distribution of proved oil reserves. Table 2–3 lists the non–U.S.-based companies with the largest estimated oil production and reserves as of late 2011.

In a 2008 study, the U.S. Geological Survey estimated that the offshore Arctic region may hold 90 billion barrels of recoverable oil—almost one-fourth of the world's unmapped reserves.[3] About 65% is thought to be in Russian territorial waters.

Royal Dutch Shell separately estimated that the Arctic also holds about 30% of the world's yet-to-be-found natural gas, according to a March 2011 news report.[4] There has been some exploration of this vast resource but no development, although several nations (including Russia, Greenland, and the United States) are jockeying for position in planning for future activity. Major oil companies are doing the same, including Shell, Statoil, ExxonMobil, Chevron, BP, and the Russian state oil company, Rosneft.

Table 2–2. Nations with largest estimated oil reserves, 2012

Country[a]	Estimated proved oil reserves as of January 1, 2012,[b] billion barrels	Producing oil wells as of December 31, 2010[c]	Estimated oil production in 2011, million b/d
Saudi Arabia	264.52	2,895	9.00
Venezuela	211.17	14,651	2.51
Canada	173.62	62,519	2.86
Iran	151.17	2,074	3.58
Iraq	143.10	1,526	2.48
Kuwait	101.50	1,286	2.18
United Arab Emirates[d]	97.80	1,458	2.50
Russia	60.00	107,476	10.32
Libya	44.10	2,060	0.44
Nigeria	37.20	2,068	2.18
Kazakhstan	30.00	1,256	1.60
Qatar	25.38	513	0.81
United States	20.68	373,648	5.60
China	20.35	71,552	4.09
Brazil[e]	13.99	11,797	2.08
Algeria	12.20	2,014	1.27
Mexico	10.16	7,476	2.54
Angola	9.50	1,321	1.64
Azerbaijan	7.00	62	0.93
Ecuador	7.21	3,447	0.50
India	8.93	3,719	0.78
Norway	5.32	812	1.74
Oman	5.50	2,286	0.89
Neutral Zone (Iraq/Saudi Arabia)	5.00	761	0.59
Sudan[f]	5.00	42	0.47

Source: *Oil & Gas Journal* survey data as of Jan. 1, 2012.

[a] Only nations with at least 5.0 billion barrels of estimated reserves are shown.

[b] All reserves are reported as proved reserves recoverable with current technology and prices except those for Russia; that figure includes proved plus some probable. Reserves data for Canada includes tar sands.

[c] Well count does not include shut-in, injection, or service wells.

[d] Abu Dhabi has 94.3% of the reserves in the seven United Arab Emirates.

[e] Brazil may have 50–100 billion barrels of additional reserves in presalt offshore formations.

[f] South Sudan gained independence from Sudan on July 9, 2011; no separate data for South Sudan were available for presentation in this book.

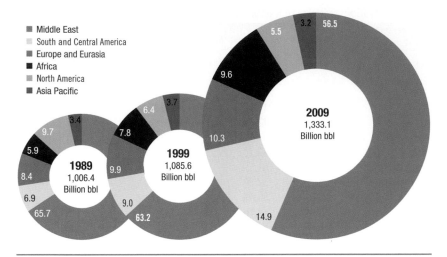

Fig. 2–1. Distribution of proved worldwide oil reserves: 1989, 1999, and 2009 (billion barrels) (*Source: BP Statistical Review of World Energy 2010*, June 2010)

Table 2–3. Top non–U.S.-based energy companies in oil production and reserves, 2011

Company	Production, million barrels	Reserves, million barrels
Saudi Arabian Oil	2,920.0	260,100.0
National Iranian Oil	1,350.5	137,010.0
Pemex	957.0	10,359.1
BP	866.0	10,530.0
Iraq National Oil	861.4	115,000.0
PetroChina	858.0	11,278.0
OAO Rosneft	847.4	18,110.0
PdVSA	814.0	211,170.0
Abu Dhabi National Oil	789.1	92,200.0
Petrobras	783.3	10,757.2
Kuwait Petroleum	741.0	101,500.0
OAO Lukoil	708.1	13,025.0
Nigerian National Petroleum	666.1	37,200.0
Sonangol	653.4	9,500.0

Source: Oil & Gas Journal, October 3, 2011, *OGJ100* survey data.

For example, Statoil expressed optimism in April 2011 about a new oil find about 125 miles offshore Norway that it said could hold as much as 500 million boe.[5] The new field is about 60 miles north of the company's Snøhvit gas field, which was the first Norwegian energy project in the Arctic. The Norwegian Petroleum Directorate estimates that the country's portion of the Barents Sea could hold as much as 6 billion boe.

In August 2011, ExxonMobil entered into a joint venture with Rosneft that included (among other terms) opportunities to explore in a Russian sector of the Arctic Ocean.[6] An initial ExxonMobil investment of $2.2 billion is envisioned for activities in the Kara Sea, between the northern coast of European Russia and the Novaya Zemlya island chain. (A similar deal, proposed earlier in the year by BP, fell through.)

The Arctic wilderness poses special risks, ranging from drifting icebergs to hostile weather. Industry experts, however, have contended that exploration, while environmentally sensitive, is not technically difficult in comparison with areas such as the deep waters of the Gulf of Mexico. They expect that drilling will be in relatively shallow waters and that there would be fewer storms in the region. Changing global climate conditions may lead to expansion of open water in the Arctic in the years ahead.

Recent Trends in U.S. Oil Production

In the United States, domestic oil production declined steadily from 1985 to 2008. However, that trend was reversed by the successful exploitation of unconventional reserves such as shale oil, using two technologies: directional drilling and hydraulic fracturing (described in chapter 5).

As of early February 2011, onshore drilling for oil had risen to almost twice the level seen one year earlier, with 818 oil-directed rigs in operation. This was the highest number since 1987, according to the oil service company Baker Hughes.

Figure 2–2 presents data from EIA on U.S. domestic oil production in recent years. Tables 2–4 and 2–5 present information on liquids reserves and liquids production, respectively, for the U.S.-based companies that comprise the top 15 in domestic reserves and domestic production.

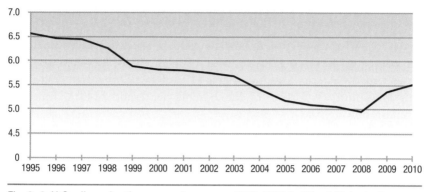

Fig. 2–2. U.S. oil production, annual average: 1995–2010 (million b/d, including crude oil and lease condensate) (*Source:* EIA)

Table 2–4. Top U.S.-based energy companies in domestic liquids reserves, 2011

Company	Liquids reserves, million barrels	
	United States	Worldwide
ExxonMobil	1,952.0	9,418.0
ConocoPhillips	1,934.0	3,616.0
Occidental Petroleum	1,697.0	2,476.0
Chevron	1,275.0	6,503.0
Apache	759.0	1,308.7
Devon Energy	597.0	1,160.0
Pioneer Natural Resources	544.9	565.0
EOG Resources	505.9	537.8
Anadarko Petroleum	498.0	749.0
Denbury Resources	338.3	338.3
Hess	304.0	1,104.0
Whiting Petroleum	254.3	254.3
Sandridge Energy	252.1	252.1
Noble Energy	225.0	365.0
Continental Resources	224.8	224.8

Source: Oil & Gas Journal, October 3, 2011, *OGJ150* survey data.

EIA also reported that, in 2010, the United States was still importing about 50% of its oil, even though domestic output of crude oil and related liquids rose by 3% compared to 2009—the highest level in almost a decade. The major reason for the increase was (as noted above) the wider use of unconventional extraction methods to exploit onshore resources.

More specifically, EIA said that domestic production of crude oil and related liquids rose to an average of 7.51 million b/d in 2010, the

Table 2–5. Top U.S.-based energy companies in domestic liquids production, 2011

Company	Annual liquids production, million barrels United States	Worldwide
Chevron	178.0	702.0
ConocoPhillips	139.0	318.0
ExxonMobil	123.0	709.0
Occidental Petroleum	99.0	201.0
Anadarko Petroleum	47.0	73.0
Devon Energy	44.0	73.0
Apache	40.3	125.1
EOG Resources	33.9	38.4
Hess	32.0	112.0
Marathon Oil	25.0	89.0
Denbury Resources	21.9	21.9
Whiting Petroleum	19.0	19.0
Noble Energy	19.0	31.0
Pioneer Natural Resources	17.5	31.8
Plains Exploration & Production	16.8	16.8

Source: Oil & Gas Journal, October 3, 2011, OGJ150 survey data.

highest since 2002. The increase made possible a 2% drop in U.S. crude oil imports, to 9.45 million b/d, despite rising demand as the economy recovered. Table 2–6 shows the nations from which the United States imported most (approximately 82%) of its crude oil in 2010, as well as preliminary crude import data for 2011. Table 2–7 provides similar information on 2010 total petroleum imports (crude plus products) for five other nations and Europe.

Table 2–6. U.S. crude oil import sources: Top 10 countries (million b/d)

Country	Crude imports to U.S., million b/d 2010	2011 (prelim)
Canada	1.970	2.213
Mexico	1.152	1.117
Saudi Arabia	1.082	1.259
Nigeria	0.983	0.757
Venezuela	0.912	0.865
Iraq	0.415	0.458
Angola	0.383	0.315
Colombia	0.338	0.403
Algeria	0.328	0.189
Ecuador	0.210	0.228

Source: U.S. DOE/EIA. Published in Oil & Gas Journal, Jan. 9, 2012, p. 31.
Note: Includes imports for the Strategic Petroleum Reserve.

Table 2–7. Major sources of petroleum imports for selected countries/regions, 2010

Imports to:	Europe	United States	Former Soviet Union	Middle East	West Africa	North Africa	South/Central America
Canada	217	124			137		
Mexico	90	477					26
Europe			5,982	2,355		1,677	
China			676	2,383	878		
India				2,612	428		193
Japan		95	293	3,629			

Source: BP Statistical Review of World Energy 2011, June 2011: Inter-area movements (oil).
Note: Figures are given in thousand b/d and include imports of crude oil as well as petroleum products.

Products Created from Oil

Because of its high energy density, easy transportability, and relative abundance, oil has been the world's most important source of energy since the mid–1950s. Refineries convert crude oil into a wide range of products. The refining processes used to achieve this conversion are described in chapter 11.

According to EIA, a 42-gallon barrel of crude oil typically yields 44 gallons of petroleum products (owing to a phenomenon called *refinery gain*). About 85% of the product yield (fig. 2–3) consists of liquid fuels, including gasoline, diesel fuel, kerosene, jet fuel, fuel oil, and propane. The 15% of the petroleum not used to create liquid fuels, according to EIA, is used as a raw material for producing a wide range of chemical products, including fertilizers, pesticides, pharmaceuticals, solvents, and plastics.

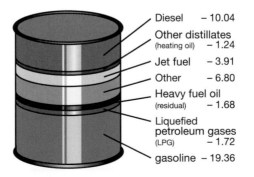

Diesel – 10.04
Other distillates (heating oil) – 1.24
Jet fuel – 3.91
Other – 6.80
Heavy fuel oil (residual) – 1.68
Liquefied petroleum gases (LPG) – 1.72
gasoline – 19.36

Fig. 2–3. Products made in the U.S. from a barrel of crude oil, 2009 (gallons) (*Source:* DOE)

Certain types of hydrocarbons produced during refining can be mixed with nonhydrocarbons to create other products, such as greases, asphalt, petroleum coke, sulfur, sulfuric acid, and wax. More details on this subject are presented in chapter 11.

Recent Trends in Oil Consumption

Global use of oil has been on a steady upward trajectory for decades, as nations around the world have used oil-based fuels to expand their industries and economies (fig. 2–4). Global demand in 2007 reached about 86 million b/d but fell in both 2008 and 2009 as a global financial crisis (triggered mainly by mortgage lending and related activities) depressed investment and dampened economic activity. The figure for 2009 was 84.1 million b/d. For 2010, DOE reported that total world oil consumption resumed its upward trend, reaching about 87.9 million b/d and setting a new record.

At the 2009 global consumption rate of about 84.1 million b/d (30.7 billion barrels per year) and using the *Oil & Gas Journal* estimate of proved oil reserves (1.35 trillion barrels), the remaining oil supply would last about 44 years. Using the DOE reserves estimate of 3.74 trillion barrels (which includes oil sands), the remaining oil supply would last about 122 years.

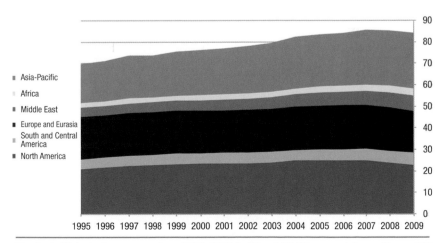

Fig. 2–4. Global oil consumption: 1995–2009 (million b/d) (*Source:* Adapted from *BP Statistical Review of World Energy 2010*, June 2010)

Looking just at the United States, about two-thirds of all oil use is for transportation, according to DOE (fig. 2–5). In other nations, oil is more commonly used for space heating and power generation than for transportation.

Totals include such products as asphalt, road oil, kerosene, LP gases, lubricants, and petroleum coke, as well as gasoline, diesel, aviation fuels, heating oil, and residual fuel oil. Other petroleum products commonly used for transportation include diesel fuel (used for trucks, buses, railroads, some vessels, and a few passenger autos), jet fuel, and residual fuel oil (used for tankers and other large vessels).

In nontransportation uses—such as space heating, industrial operations, and power generation—substitution of other energy sources for oil has been much easier than in the transportation sector. As a result, U.S. nontransportation use of oil fell from a peak of 8.7 million b/d in 1978 (about 47% of total oil use) to a low of less than 6 million b/d in the late 1980s and early 1990s. Consumption in these sectors has risen to 6.5–7.0 million b/d more recently. Oil accounts for less than 20% of the energy consumed for stationary uses, down from 30% in 1973.

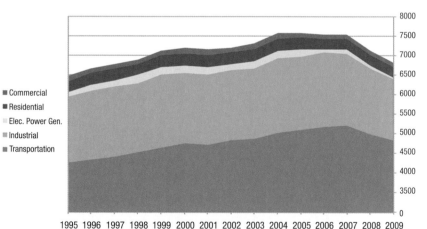

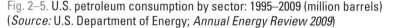

Fig. 2–5. U.S. petroleum consumption by sector: 1995–2009 (million barrels) (*Source:* U.S. Department of Energy; *Annual Energy Review 2009*)

Projected Trends in Oil Consumption

In July 2010, DOE published its *International Energy Outlook 2010*. The study's reference case scenario projects that total global consumption of liquid fuels will grow to 110.6 million b/d in 2035 as world economic growth resumes (fig. 2–6). DOE expects an annual growth rate of 3.4% in world gross domestic product over its projection period of 2007–2035. DOE includes as liquid fuels not only products derived from conventional resources (chiefly crude oil) but also products derived from unconventional sources (biofuels, oil sands, extra-heavy oil, coal-to-liquids, gas-to-liquids, and shale oil) and natural gas liquids.

In an early release (December 2010) of its *Annual Energy Outlook* for 2011, EIA said it expected increased oil production from the members of OPEC, as well as Russia, Kazakhstan, Brazil, and Canada. The agency also said it expected a moderate increase (in real terms) in the price of oil, from about $88 per barrel in mid–December 2010 to $125 in 2009 dollars (about $200 in nominal terms) by 2035.

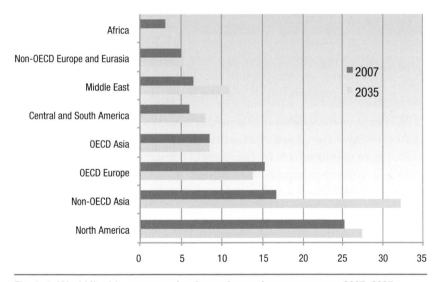

Fig. 2–6. World liquids consumption by region and country group: 2007–2035 (million b/d) (*Source:* DOE, *International Energy Outlook 2010*)

Another agency that produces well regarded forward-looking views of world oil consumption is the International Energy Agency (IEA), based in Paris. IEA released its *World Energy Outlook* in November 2010. Its central analysis—the New Policies Scenario (NPS)—assumes that nations around the world take some steps to address concerns about global climate change, but not the aggressive actions urged by IEA itself. The NPS makes the following projections concerning oil use through the year 2035:

- Oil demand (excluding biofuels) reaches 99 million b/d by 2035. All of the net growth comes from non-OECD countries (almost half from China alone). OECD is the Organization for Economic Cooperation and Development.

- Crude oil production reaches about 68 million b/d by 2020 but never regains its all-time peak level of 70 million b/d reached in 2006.

- Oil from unconventional sources (e.g., shale, tar sands, and extraheavy oil) meets about 10% of total world oil demand by 2035 (compared to 3% in 2009). Canadian oil-sands production reaches 4.2 million b/d in 2035 (compared to 1.3 million b/d in 2009).

- Oil production from the Caspian Sea region grows from 2.9 million b/d in 2009 to about 5.2 million b/d in 2035. Financing and multination pipeline construction challenges could constrain this development.

With regard to the United States, EIA issued its *Annual Energy Outlook 2011 (AEO2011)* in April 2011, examining U.S. energy trends through 2035. In the study's reference case, total U.S. consumption of liquid fuels, including both fossil liquids and biofuels, grows from 18.6 million b/d (36.2 quadrillion Btu [quads] per year) in 2009 to 21.4 million b/d (41.7 quads per year) in 2035 (fig. 2–7).

The transportation sector dominates demand for liquid fuels in *AEO2011*, but its share (as measured by energy content) grows only slightly—from 72% of total liquids consumption in 2009 to 74% in 2035. Biofuel consumption accounts for most of the growth. The biofuel portion of liquid fuels consumption is about 1.89 million b/d equivalent (3.9 quads per year) in 2035. Oil consumption is expected to remain roughly unchanged by 2035, at about 20 million b/d.

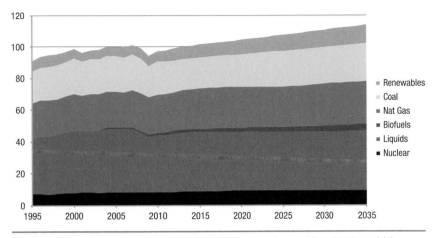

Fig. 2–7. U.S. primary energy use by fuel: 1995–2035 (quadrillion Btu/year) (*Source:* EIA, *AEO2011*)

References

1 Oil Sands InfoMine. http://oilsands.infomine.com/countries/ (accessed February 21, 2011).

2 Bureau of Land Management, Colorado State Office, Lakewood. Oil shale and tar sands EIS information center. Oil shale/tar sands guide: About oil shale. http://ostseis.anl.gov/guide/oilshale/index.cfm (accessed February 21, 2011).

3 U.S. Geological Survey. 2008. 90 Billion barrels of oil and 1,670 trillion cubic feet of natural gas assessed in the Arctic. News release, July 23.

4 Pfeifer, S. 2011. Arctic frontier: Huge prize lies under a pristine wilderness. *Financial Times*, March 21.

5 Ward, A. 2011. Statoil looks to Arctic find to boost prospects. *Financial Times*, April 10.

6 Kramer, A. E. 2011. Exxon wins prized access to Arctic with Russia deal. *New York Times*, August 30.

3 | Natural Gas Overview

This chapter provides information about the composition of natural gas, the location of major gas resources around the world, and current patterns of, as well as future projections for, gas production and use.

Formation and Composition of Natural Gas

The simplest hydrocarbon molecule—one carbon atom bonded to four hydrogen atoms (CH_4)—is called *methane*. It is the primary component of natural gas.

In addition to methane, raw natural gas may also contain other hydrocarbons (see table 3–1), as well as water, carbon dioxide, oxygen, nitrogen, hydrogen sulfide, and even helium. (Gas from the Hugoton field in West Texas, Oklahoma, and Kansas contains as much as 2% helium, a light gas used in electronics manufacturing.)

The nonhydrocarbons (sometimes called *inerts*) must be removed from raw natural gas. Some can be recovered and sold as by-products. (Chap. 7 describes the removal of inerts; chap. 12 describes the extraction of nonmethane hydrocarbons.)

Table 3–1. Typical hydrocarbon composition of raw natural gas

Hydrocarbon	Composition, %
Methane	70–98
Ethane	1–10
Propane	Trace to 5
Butane	Trace to 2

Source: Hyne, N. J.

Natural gas has no odor or color and is lighter than air. An organic sulfur compound called *mercaptan* is added to natural gas to give the gas its identifiable smell (see chap. 9).

Because natural gas has a specific gravity just over half that of air, when released into the atmosphere, it rises and quickly mixes with air. Gases that are heavier than air—such as propane, the vapors of gasoline, and fuel oils—will tend to pool or settle at ground level.

As noted in chapter 1, the source of both oil and natural gas is organic matter buried in layers of sedimentary rocks created over eons (fig. 3–1). Black shale is the kind of rock most commonly rich in organics, deposited as mud on the floor of ancient oceans. The organic matter changes to oil and gas under the influence of temperature and pressure, and a portion of the hydrocarbons migrate through the rock to eventually collect in geologic structures called *traps*.

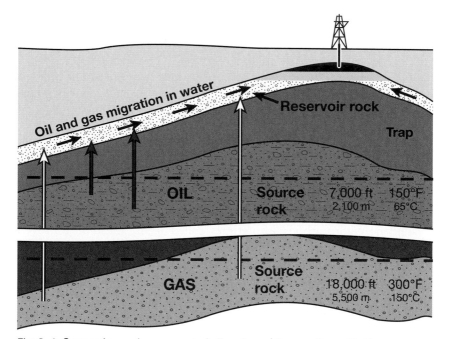

Fig. 3–1. Generation and movement of oil and gas (*Source:* Hyne, N. J.)

Natural gas also may contain heavier hydrocarbons in the gaseous state, such as pentane, hexane, and heptane. At surface conditions, these will condense out of the gas to form natural gas condensate, often shortened to *condensate*. Condensate resembles gasoline in appearance and is similar in composition to some volatile light crude oils.

Conventional Natural Gas

The term *conventional* is generally used to describe natural gas produced from well-understood geologic formations known through experience to hold natural gas. Examples include limestones and sandstones at depths of a few thousand feet (fig. 3–2).

In North America, most conventional natural gas has been located in relatively distinct geographic areas (basins). The greatest U.S. natural gas reserves historically have been concentrated around Texas and the Gulf of Mexico, with substantial amounts also found in the Rocky Mountain West (e.g., the Green River and Piceance basins) and Alaska.

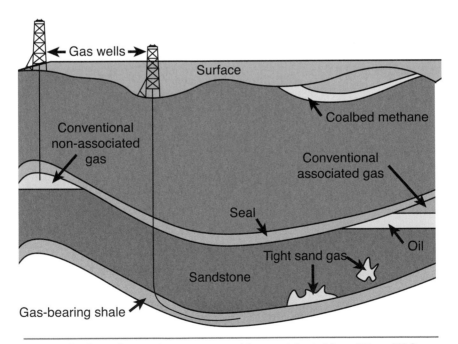

Fig. 3–2. Geology of natural gas resources (*Source:* Adapted from EIA and U.S. Geological Survey)

Conventional gas that is produced during the extraction of crude oil is commonly referred to as *associated gas*. By contrast, a formation targeted specifically for extraction of natural gas is said to yield *nonassociated gas*.

Unconventional Natural Gas

In the broadest sense, *unconventional* natural gas is gas that cannot be economically produced unless one or more technologies are used to stimulate the gas-bearing formation and to expose more of the formation to the wellbore. An unconventional gas reservoir can be deep or shallow, of high or low pressure and temperature, and can contain single or multiple layers.

What was unconventional yesterday can nevertheless become conventional tomorrow, as technology and geologic knowledge advance and as public policy changes (e.g., tax incentives to promote investment). For example, before 1978, unconventional natural gas deep in the Anadarko basin (centered in western Oklahoma and the Texas Panhandle) was virtually untouched, because it was neither economical nor technically possible to extract it. However, regulatory changes and passage of the Natural Gas Policy Act of 1978 provided incentives for searching for and extracting unconventional natural gas and spurred investment into deep exploration and development drilling. These changes made much of the deep gas in the basin conventionally extractable.

Today, there are six main categories of unconventional natural gas:

- Deep gas
- Tight gas
- Shale gas
- Coal-bed methane
- Gas in geopressurized zones
- Methane hydrate

Deep gas

Deep gas is typically found 15,000 feet or deeper underground, considerably deeper than conventional gas. Therefore, deep gas is relatively expensive to find and produce.

Tight gas

Tight gas is trapped in unusually impermeable hard rock or in sandstone or limestone that is highly nonporous (tight sand). Extraction of gas from tight formations typically requires expensive techniques such as fracturing and acidizing.

Shale gas

Natural gas can also exist in deposits of shale, a fine-grained and soft sedimentary rock that breaks easily into thin, parallel layers. Gas is typically found in sections where two thick, black shale deposits "sandwich" a thinner area of shale. This gas can be held in naturally occurring fractures or in pore spaces, or it can be adsorbed onto the surface of the organic components of the shale.

Ordinarily, shales are not permeable enough to let significant amounts of gas or other fluids flow to a wellbore; for this reason, most shales are not commercially viable sources of natural gas. However, over the past 20 or so years, considerable improvements in technology—notably, directional and horizontal drilling and multistage hydraulic fracturing—have contributed toward bringing shale gas into the U.S. energy portfolio. Horizontally drilled wellbores—some as long as 10,000 feet—maximize the amount of borehole surface area in contact with the shale, while fracturing (as explained in chap. 5) creates artificial fissures in the shale that facilitate fluid flow from the formation into the wellbore.

Major gas shale plays in the United States are the Marcellus (Appalachia), Haynesville (Louisiana/Texas), Barnett (Texas), Fayetteville (Arkansas), Woodford (Oklahoma), Eagle Ford (Texas), and Antrim (Michigan). Major plays in Canada are the Horn River/Muskwa (northeast British Columbia), Colorado Group (west-central Saskatchewan/south-central Alberta), and Montney (Alberta).

As of 2008, the Barnett shale was already the source of about 6% of all natural gas produced in the Lower 48 states. More details on the extent of gas-bearing shales in the United States and in other countries are presented later in this chapter (in the "Resources and Reserves" section).

Productivity of shale gas wells

There is an emerging debate about how rapidly production from a typical shale gas well is likely to decline and whether that decline rate flattens out over time.

If the curve flattens out slowly, then shale gas wells will produce gas at a reasonably high rate and low cost over a long period of time. By contrast, if the curve declines more sharply and rapidly, then the ultimately recovered gas reserves from a shale formation could be significantly lower and the production costs could be higher.

For this reason, some experts argue that better models are needed to guide the commitment of financial resources to the continued development of gas-bearing shales.

Coal-bed methane

Historically, methane was a nuisance and a safety hazard in the coal-mining industry. As a coal mine is being built and coal extracted, methane held within the coal leaks out into the mine. This poses a safety threat because too high a concentration of methane creates dangerous conditions for coal miners. Thus, in the past, accumulating methane was intentionally vented to the atmosphere.

Today, however, coal-bed methane has become a popular form of unconventional natural gas, with many projects put in place around the world to extract and market it. In April 2011, the Potential Gas Committee (PGC) estimated that about 158.6 trillion cubic feet (tcf) of technically recoverable (probable, possible, and speculative) coal-bed methane existed in the United States at the end of 2010.[1] (Based at the Colorado School of Mines for more than 40 years, the nonprofit PGC assesses U.S. natural gas resources annually.) In eastern Australia, coal-bed gas reserves (proved plus probable) grew from about 6 tcf in 2006 to about 27 tcf in 2010, according to Resource and Land Management Services, a consultancy based in Brisbane.

Geopressurized gas

Natural geologic formations in which the pressure is higher than would be expected for a given depth are said to be *geopressurized*. In these zones, layers of clay have been deposited and compacted quickly on top of sand or silt. The water and natural gas present in the clay have been squeezed out by the compression of the clay and have entered the more porous sand or silt deposits. Owing to this compression, the natural gas in the sand or silt is under very high pressure.

Geopressurized zones are typically quite deep, usually 10,000–25,000 feet below the surface of the earth. In the United States, they are located chiefly in the Gulf Coast region. It has been estimated that U.S. geopressurized zones could hold anywhere from 5,000 to 49,000 tcf of natural gas.

Methane hydrate

Another unconventional gas resource under evaluation for potential production in the longer term is methane hydrate. *Hydrate* is typically a cold, slushlike, crystalline structure consisting of methane molecules trapped in a lattice of water molecules. Such hydrates are abundant in the Arctic (where they were first discovered) and in marine sediments, below the seabed.

Estimates of the worldwide methane hydrate resource vary from 7,000 to more than 73,000 tcf; however, these numbers are far from certain. Substantial additional research is needed to assess how hydrates form, interact with surrounding materials, behave during extraction operations, and potentially affect the environment.

In summary, unconventional natural gas constitutes a large proportion of the natural gas that is left to be extracted in North America and is playing an ever-increasing role in supplementing the U.S. natural gas supply. A study conducted by ICF International concluded that, in 2007, abut 42% of total U.S. natural gas production was from unconventional sources, and this figure is expected to rise to 64% in 2020.[2]

Unconventional natural gas also will play a major role in global energy markets. According to EIA, the full extent of the world's tight gas, shale gas, and coal-bed methane resource base remains to be assessed fully. However, EIA projects a substantial increase in those supplies—especially in the United States, as well as in Canada and China.

EIA's *International Energy Outlook 2010* projects that shale gas alone could account for 26% of U.S. natural gas production in 2035. According to DOE, in Canada and China, tight gas, shale gas, and coal-bed methane are expected to account for 63% and 56% of total domestic production, respectively, in 2035.[3]

Liquefied natural gas (LNG)

In most discussions about natural gas, reference is made to liquefied natural gas (LNG). Importantly, LNG is not a naturally occurring gas resource. Rather, it is created by deeply cooling natural gas (typically to 260°F below zero).

The process transforms natural gas from gaseous to liquid form—increasing its energy density (Btu per cubic foot) by a factor of about 600—and makes possible the cost-effective shipment of the LNG in insulated tanker ships from gas production regions to major market regions.

The LNG is then unloaded from the tankers and regasified (i.e., warmed to convert it back into the gaseous state), treated to bring its composition into compliance with market specifications, and then distributed to users through conventional pipeline networks.

More details about LNG are presented in chapter 10.

Resources and Reserves

A big story in 2011 concerning global natural gas resources was the growing role of gas from unconventional resources—particularly the emergence of shale gas. In North America, the number of states with significant shale gas resources has increased. States with major shale gas potential include New York, Pennsylvania, Arkansas, and Oklahoma, as well as Texas and Louisiana. For example, in August 2011, the U.S. Geological Survey estimated that the Marcellus shale (underlying several Appalachian states) contains 84 tcf of gas and 3.4 billion barrels of natural gas liquids (undiscovered and technically recoverable). In Canada, shale gas developments are growing in the Horn River area of northeastern British Columbia, and prospective shale targets have been identified in Alberta, Saskatchewan, Ontario, Quebec, New Brunswick, and Nova Scotia.

Looking more widely, EIA has identified 48 shale basins with significant gas-bearing potential in more than 30 countries—from Australia, China, and India to eight European nations. At present, these resources are in various stages of assessment and development (table 3–2).[4]

Table 3–2. Estimated shale gas technically recoverable resources for selected basins in 32 countries

Region/country	Technically recoverable shale gas resources, tcf
Europe:	
France	180
Germany	8
Netherlands	17
Norway	83
United Kingdom	20
Denmark	23
Sweden	41
Poland	187
Turkey	15
Ukraine	42
Lithuania	4
Other[a]	19
North America:	
United States	862
Canada	388
Mexico	681
Asia:	
China	1,275
India	63
Pakistan	51
Australia	396
Africa:	
South Africa	485
Libya	290
Tunisia	18
Algeria	231
Morocco	11
Western Sahara	7
South America:	
Venezuela	11
Colombia	19
Argentina	774
Brazil	226
Chile	64
Uruguay	21
Paraguay	62
Bolivia	48
Total	6,622

Source: EIA
[a] Includes Romania, Hungary, and Bulgaria.

In addition, a study released in May 2011 by the European Centre for Energy and Resource Security concluded that unconventional gas resources—notably shale gas—might "in theory . . . be able to cover European gas demand for at least another 60 years." [5] Therefore, unconventional gas resources in Europe could reduce that region's dependence on Russia and the Middle East.

With regard to the United States, EIA reported in April 2011 that U.S. proved (recoverable) reserves of natural gas (from all sources) are much higher than has previously been estimated—about 284 tcf in 2009 (the highest level since 1971).[6] That figure (which includes natural-gas-plant liquids) reflects major improvements in shale gas exploration and production technologies.

In addition to proved reserves, there are large volumes of natural gas classified as undiscovered technically recoverable resources. Combining these two classifications of resources, EIA estimates that the United States possesses 2,552 tcf of potential natural gas resources. At the 2009 rate of consumption, that is enough gas to meet U.S. demand for about 110 years. EIA also reported that technically recoverable shale gas accounts for 827 tcf of this resource estimate, more than double the 353 tcf estimated in 2010.

Table 3–3 lists those nations with significant estimated natural gas reserves, as of January 1, 2012. Figure 3-3 provides another perspective on global distribution of proved natural gas reserves.

With regard to United States, tables 3–4 and 3–5 present information on gas reserves and gas production, respectively, for the U.S.-based companies that comprise the top 15 in domestic reserves and domestic production.

Table 3–3. Nations with largest estimated natural gas reserves, 2012

Country[a]	Estimated proved gas reserves as of January 1, 2012,[b]tcf
Russia	1,680.0
Iran	1,168.0
Qatar	890.0
Saudi Arabia	283.0
United States	272.5
Turkmenistan	265.0
Abu Dhabi[c]	200.0
Venezuela	195.1
Nigeria	180.5
Algeria	159.0
Indonesia	141.1
Iraq	111.5
China	107.0
Kazakhstan	85.0
Malaysia	83.0
Egypt	77.2
Norway	70.9
Uzbekistan	65.0
Kuwait	63.0
Canada	61.0
Libya	52.8

Source: *Oil & Gas Journal*, January 1, 2012, survey data.

[a] Only nations with at least 50 tcf of estimated reserves are shown.

[b] All reserves are reported as proved reserves recoverable with current technology and prices except for Russia and Canada, for which figures includes proved plus some probable.

c Abu Dhabi has 93.0% of the gas reserves of the seven United Arab Emirates.

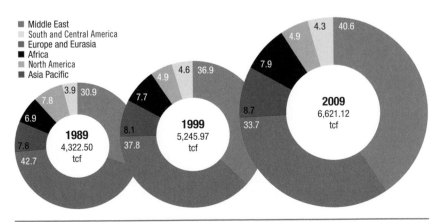

Fig. 3–3. Distribution of proved worldwide natural gas reserves (in %): 1989, 1999, and 2009 (*Source: BP Statistical Review of World Energy 2011*, June 2011)

Table 3–4. Top 15 U.S.-based energy companies in domestic natural gas reserves, 2011

Company	Gas reserves, billion cubic feet (bcf)	
	United States	Worldwide
ExxonMobil	25,994.0	46,813.0
Chesapeake Energy	11,327.0	11,327.0
ConocoPhillips	10,479.0	18,235.0
Devon Energy	9,065.0	10,283.0
Anadarko Petroleum	8,117.0	8,117.0
EOG Resources	6,491.5	8,470.2
EQT Production	5,205.7	5,205.7
Southwestern Energy	4,930.0	4,930.0
Williams Companies	4,272.0	4,272.0
Ultra Petroleum	4,200.2	4,200.2
Consol Energy	3,731.6	3,731.6
Range Resources	3,566.5	3,566.5
Apache	3,273.0	9,867.2
Petrohawk Energy	3,110.1	3,110.0
Occidental Petroleum	3,034.0	5,320.0

Source: Oil & Gas Journal, October 3, 2011, OGJ150 survey data.

Table 3–5. Top 15 U.S.-based energy companies in domestic natural gas production, 2011

Company	Annual gas production, bcf	
	United States	Worldwide
ExxonMobil	1,057.0	2,920.0
Anadarko Petroleum	829.0	829.0
Chesapeake Energy	775.0	775.0
ConocoPhillps	764.0	1,794.0
Devon Energy	716.0	930.0
Chevron	479.0	1,839.0
EOG Resources	422.6	633.4
Williams Companies	420.0	420.0
Southwestern Energy	403.6	403.6
Apache	266.8	689.4
Occidental Petroleum	247.0	431.0
Petrohawk Energy	234.5	234.5
El Paso	216.0	226.0
Ultra Petroleum	205.6	205.6
QEP Resources	203.8	203.8

Source: Oil & Gas Journal, October 3, 2011, OGJ150 survey data.

Recent Trends in Gas Production

Figure 3–4 shows the trend in global natural gas production from 1995 to 2009, based on data compiled by BP.[7] Following more than a decade of steady increase, gas production fell by 2.1% in 2009, the first decline on record. Production fell sharply in Russia (by 2.62 tcf) and Turkmenistan (by 1.05 tcf)—in each case, the largest decline on record. The global decline reflected the impact of the worldwide recession that began in late 2008. (Gas production in the United States actually increased in every year from 2007 to 2009, as noted below.)

As shown in figure 3–5, natural gas production in the United States grew from 1995 to 2001, then fell back to about the 1995 level for 2004 to 2005. However, in 2006, growth resumed for three straight years, and production reached just under 21 tcf in 2009. Driving that growth of 2006–2009 was the increasing contribution from gas shales (coupled with continued strong production of gas from tight formations). The combination of horizontal drilling and hydraulic fracturing technologies has made it possible to produce shale gas economically, leading to an average annual growth rate in production from that resource of 48% from 2006 to 2010.[8]

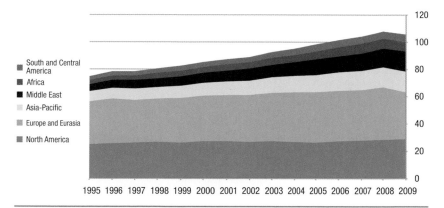

Fig. 3–4. Annual global natural gas production (in tcf), 1995–2009 (*Source: BP Statistical Review of World Energy 2010*, June 2010)

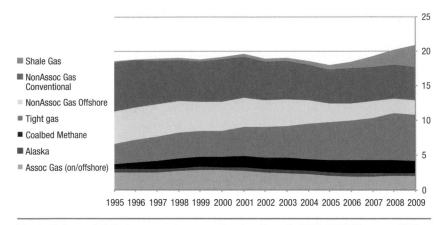

Fig. 3–5. Annual U.S. dry natural gas production (in tcf), 1995–2009 (*Source:* EIA)

ProjectedTrends in Gas Production

Looking to the future, gas resources from a variety of sources are expected to keep world markets well supplied at relatively low prices for the period from 2007 to 2035, according to EIA.[9] The major increase in production is expected from the Middle East, Africa, and Russia (fig. 3–6). EIA projects that Iran and Qatar alone will increase production by a combined 12 tcf by 2035—nearly one-fourth of the total growth in global production.

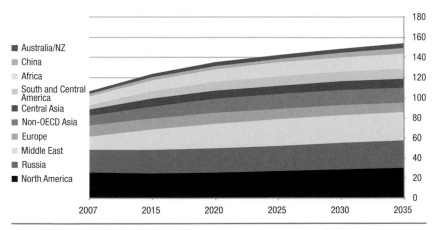

Fig. 3–6. Projected annual global natural gas production (in tcf), 2007–2035 (*Source:* EIA, *International Energy Outlook 2010*)

Tight sands, gas-bearing shales, and coal beds are expected to make major contributions to growth in gas production—and not just in the United States. For example, these three resources are expected to account for 63% of total domestic production in Canada and 56% in China in 2035.

In the United States, total domestic natural gas production is projected to grow from about 21 tcf in 2009 to 26.3 tcf in 2035 (fig. 3–7), with shale gas more than offsetting a decline in production of conventional natural gas. Shale gas production is expected to grow more than fivefold over the analysis period, reaching 12.2 tcf to account for 46% of all domestic U.S. gas production in 2035.

As noted earlier, EIA estimates U.S. technically recoverable unproved shale gas resources at 827 tcf. Although more information has become available as a result of increased drilling activity in developing shale gas plays, EIA has cautioned that estimates of technically recoverable resources and well productivity remain highly uncertain. Also note that gas from tight formations is expected to be a major contributor to U.S. production through 2035.

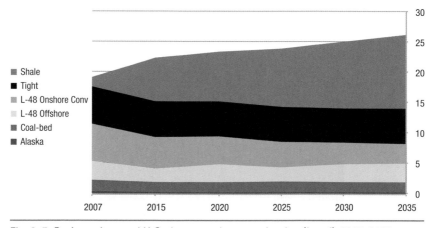

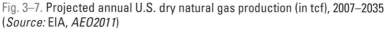

Fig. 3–7. Projected annual U.S. dry natural gas production (in tcf), 2007–2035 (*Source:* EIA, *AEO2011*)

Recent and Projected Trends in Gas Consumption

From 1995 to 2008, global demand for natural gas grew steadily, from about 96 tcf to about 124 tcf. But in 2009, as the global recession began to affect international energy markets, world consumption fell back to about 120 tcf, according to analysis by BP (fig. 3–8).[10]

As recovery from the economic downturn began, global demand was expected to grow by about 1.8% per year beginning in 2007, continuing at that rate through 2020. From 2020 to 2035, the growth in consumption slows to about 0.9 per year as increasingly expensive natural gas is brought to market.[11]

Over the period from 1997 to 2009, annual natural gas demand in the United States fluctuated within a remarkably small range (fig. 3–9), between 22 and 23 tcf. Minor dips in 2001 and 2009 occurred in reaction to economic downturns in those years.

Looking ahead, EIA has projected that global gas demand will climb by 44% between 2009 and 2035 (from 108.5 tcf to 156.3 tcf), with the most significant portion of that growth occurring in non–OECD nations (fig. 3–10).[12] China will lead the way in gas demand growth, accounting for more than 20% of the global increase through 2035, according to EIA.

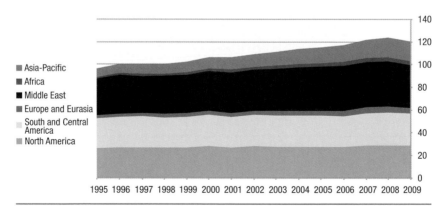

Fig. 3–8. Annual global natural gas consumption (in tcf), 1995–2009 (*Source:* BP *Statistical Review of World Energy 2010*, June 2010)

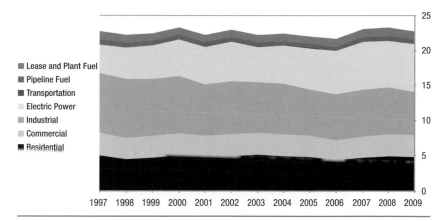

Fig. 3–9. Annual U.S. natural gas consumption (in tcf), 1997–2009 (*Source: BP Statistical Review of World Energy 2010*, June 2010)

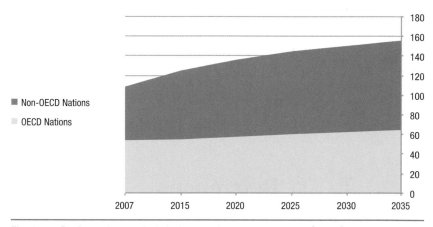

Fig. 3–10. Projected annual global natural gas consumption (in tcf), 2007–2035 (*Source:* EIA, *International Energy Outlook 2010*)

EIA also estimated that the rate of growth in U.S. natural gas consumption from 2020 to 2035 is expected to average 0.9% per year as prices rise and increasingly expensive gas resources are brought to market. In a similar U.S.-focused evaluation, EIA projected that annual U.S. gas use will rise from 22.7 tcf in 2009 to 26.5 tcf in 2035, (fig. 3–11), with natural gas thus continuing to provide about 25% of total U.S. primary energy needs over that period.[13]

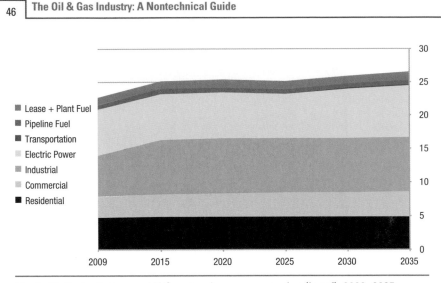

Fig. 3–11. Projected annual U.S. natural gas consumption (in tcf), 2009–2035 (*Source:* EIA, *Annual Energy Outlook 2011*)

In a look at world energy trends in 2010, the Paris-based IEA—founded in response to the 1973–74 Arab oil embargo—has summarized the future for natural gas as follows: "Natural gas is set to play a central role in meeting the world's energy needs for at least the next two-and-a-half decades."[14] The fuel's favorable environmental and practical attributes, plus expected constraints on how quickly low-carbon energy technologies can be deployed, are seen as major drivers for this trend.

References

1 Potential Gas Committee, Colorado School of Mines. 2011. Potential Gas Committee reports substantial increase in magnitude of U.S. natural gas resource base. News release, April 27. Golden, CO. http://oilsands.infomine.com/countries/ (accessed February 21, 2011).

2 ICF International. 2008. *Unconventional Natural Gas: Availability, Economics and Production Potential of North American Unconventional Natural Gas Supplies.* Fairfax, VA: ICF International. For the INGAA Foundation, Inc., Washington, DC.

3   U.S. Department of Energy, Energy Information Administration. 2010. *International Energy Outlook 2010.* Washington, DC: U.S. Department of Energy.

4   U.S. Department of Energy, Energy Information Administration. 2011. *World Shale Gas Resources: An Initial Assessment of 14 Regions Outside the United States.* Washington, DC: U.S. Department of Energy.

5   Kuhn, M., and F. Umbach. 2011, *Strategic perspectives of unconventional gas: A game changer with implications for the EU's energy security.* London. European Centre for Energy and Resource Security (EUCERS) at King's College.

6   U.S. Department of Energy, Energy Information Administration. 2011. *Annual Energy Outlook 2011.* Washington, DC: Department of Energy.

7   BP. *Statistical Review of World Energy 2010.* June 2010. London.

8   EIA. *Annual Energy Outlook 2011.*

9   EIA. *International Energy Outlook 2010.*

10   BP. *Statistical Review of World Energy 2010.*

11   EIA. *International Energy Outlook 2010.*

12   Ibid.

13   EIA. *Annual Energy Outlook 2011.*

14   International Energy Agency. 2010. *World Energy Outlook: 2010 Edition.* Paris.

4

Searching for and Evaluating Oil and Gas

Firms engaged in oil and gas exploration and development (E&D) face a daunting task. Drilling a single well on land can cost hundreds of thousands of dollars. Even more daunting, the capital investment needed to develop a medium-sized offshore field (e.g., with 100 million barrels of recoverable reserves) is on the order of $1 billion, depending on water depth, well depth, and geologic parameters.

Based on many decades of experience (their own and that of others), firms have identified in general terms those regions in which important oil-bearing formations may be found. In addition, they can use a variety of technical tools and procedures (described below) to pinpoint and analyze more accurately hydrocarbon deposits in a specific field or formation that shows promise. Beyond these geologic parameters and the related technical assessment, how does a company determine just where and when to undertake its next exploratory or developmental drilling program?

This chapter provides information on the technical assessment of potential hydrocarbon-bearing formations, the legal and contractual aspects of obtaining the right to explore for and develop such formations, and the economic and risk-assessment methods used by the petroleum industry to guide decision-making.

Obtaining Rights to Explore for and Produce Oil and Gas

The ownership or control of an area of interest (onshore or offshore) may rest with a government, a business entity, or an individual. For that reason, an E&D firm must enter into a business arrangement that will define the rights and obligations of all parties before exploratory drilling begins. In view of the risks involved, oil companies often establish joint ventures or joint operating agreements, with one firm in the partnership designated as the operator who actually supervises the work.

In the United States, most onshore oil, gas, and mineral (OGM) rights are owned by private individuals. Sometimes this is not the same person who owns the surface rights. In this case, a company must negotiate terms for a lease with the individual who owns the OGM rights.

Exploration of offshore and remote areas is generally undertaken only by very large corporations or national governments. In most nations, the government issues licenses to explore, develop, and produce its oil and gas resources. The licenses are typically administered by an oil ministry. Discussion here will focus on cases in which a government owns or controls the region of interest (and, by implication, the hydrocarbon resources that may lie beneath it).

Exploration and production licenses are awarded in competitive bid rounds and are typically based on the size of the work program proposed (e.g., the number of wells to be drilled or the level of seismic survey activity undertaken). Two kinds of license arrangements are commonly used by the oil and gas industry around the world: production-sharing agreements and service or production contracts.

Production-sharing agreement

The production-sharing agreement (PSA; also called a production-sharing contract [PSC]) is the most widely used business arrangement. Under a PSA, a government awards to a company the right to explore for and to produce hydrocarbons. The oil company bears the technical and financial risk of the initiative as it undertakes exploration, development, and ultimately production. If its efforts are successful, the company is allowed to use the money from sale of the produced oil to recover capital and operational expenses. The remainder is split between the government and the company according to a predetermined ratio. Under some PSAs, changes in international oil and gas prices or production rates can affect the share of production awarded to the company.

PSAs can be of benefit to governments that lack the expertise or capital to develop their resources and wish to attract foreign companies to do so. While they can be quite profitable for the companies involved, PSAs often involve considerable risk.

Service contract

Under a service contract, an E&D company acts as a contractor for the host government and is paid to produce the hydrocarbons.

Production contract

Under a production contract, a company takes over an existing or underdeveloped field, works to improve production, and is paid an amount based on an agreed-to portion of the increased production. In some cases, a government will claim a royalty interest on any oil that a company produces. This gives the government the right to collect a stream of future payments, typically a percentage of the value of the oil produced. In addition, the government may impose taxes on profits realized by the company as a result of its oil production.

A government also may demand payment of various bonuses and fees. One example is a signature bonus, to be paid by the company at the start of the license.

Exploration Activities

Once exploration rights have been established and related business issues negotiated, the technical work of finding and evaluating promising oil- and gas-bearing formations can begin. In decades past, companies would look for evidence of seepage at the surface, indicating the presence of hydrocarbons below the surface (fig. 4–1). As noted by API, "their tools were simple, their wells were shallow, and luck was a big part of the game."[1] Today, exploration and production (E&P) firms use a variety of advanced tools and techniques to gain a detailed picture of the subsurface.

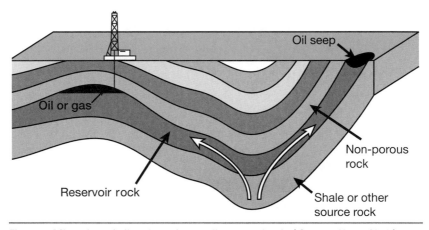

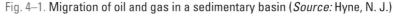

Fig. 4–1. Migration of oil and gas in a sedimentary basin (*Source:* Hyne, N. J.)

Geologic methods

Geologic techniques involve mapping and sampling of rock formations that outcrop at the earth's surface. Geologic information about an area is sometimes available from geologic agencies, based on earlier drilling activities for such resources as water, brine, coal, and minerals. These data can provide clues about the fluid content, porosity, permeability, age, and formation sequence (layering) of subsurface rocks.

Geochemical methods

Visible surface features—oil or natural gas seeps on the ground, or specific geologic features on the seafloor—sometimes indicate the presence of deep or shallow hydrocarbon deposits. Geochemical methods can evaluate the chemical and bacterial properties of the soil above suspected oil and gas reservoirs, looking for changes induced by the slow escape and upward migration of hydrocarbons.

Geophysical methods

Geophysical analysis can help to determine subsurface strata depth, thickness, and rock properties. These geophysical procedures include gravimetry, magnetometry, and seismography. (More details about each method are provided in the sections below.)

Seismographic techniques are widely used because they yield the most useful information about rock structures. In particular, they can identify traps capable of containing oil and gas. (Traps are described in detail in chap. 1.)

Survey Methods

Gravimetric survey

In a gravimetric survey, geophysicists measure variations in gravitational force. These variations give clues about the properties and extent of subsurface structures.

Magnetic survey

A magnetic survey measures minute variations in the strength and direction of the earth's magnetic field. This can also provide information about subsurface rock structures.

Seismic survey

Seismographic or seismic surveys create and then study shock waves as they are refracted (bent) and reflected by subsurface rock interfaces. Seismic methods can routinely assess structures at depths up to 20,000 feet, with an accuracy of 10–20 feet. Instruments called *geophones*, or *seismometers*, are used to detect the shock waves (fig. 4–2).

Passive seismic survey. A passive seismic survey detects natural, low-frequency movements (up to 10 cycles per second) of the earth's crust. Geophones are placed at multiple measurement points, typically several hundred yards apart, to listen for periods lasting from several hours to several days.

Active seismic survey. In an active (or seismic reflection) survey conducted on land, researchers direct sound waves into the subsurface by using large, truck-mounted mechanical *thumpers* (fig. 4–3) or by setting off small, controlled explosions.

Fig. 4–2. Low-frequency geophone (*Source:* Hyne, N. J.)

Fig. 4–3. Seismic survey vibrator truck (*Source:* Hyne, N. J.)

The sound waves are reflected from subsurface structures, traveling at different speeds and angles through rock of varying densities. The reflected sound waves are then sensed by geophones laid out in a precise pattern on the surface of the ground (fig. 4–4). Signals from the geophones are then processed by computers to create various images of the subsurface to be assessed by seismic analysts (e.g., see fig. 4–5).

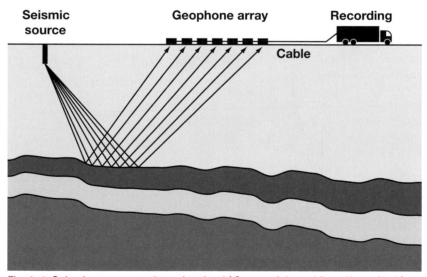

Fig. 4–4. Seismic survey conducted on land (*Source:* Adapted from Hyne, N. J.)

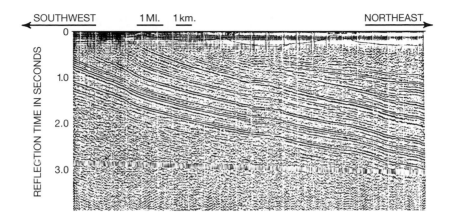

Fig. 4–5. Seismic record from Wind River basin, Wyoming (*Source:* Hyne, N. J.)

Offshore, an active seismic survey is conducted by a specially designed vessel that tows two kinds of equipment as it moves slowly across the water (fig. 4–6). The first type is a sound source (typically, an air gun), mounted close to the stern, that directs sound energy toward the ocean floor. The second type is a long plastic tube, called a *streamer*, that can stretch for up to five miles behind the vessel.

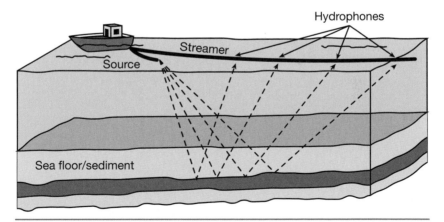

Fig. 4–6. Seismic survey conducted offshore (*Source:* Hyne, N. J.)

A streamer is designed to float 20–50 feet below the water's surface and to hold 200–300 carefully spaced vibration detectors, called *hydrophones*. To cover a large survey area, some seismic vessels can tow 12 parallel streamers and four sound sources (fig. 4–7). As in a land-based survey, signals from the hydrophones are used to create images of subsea formations.

In cases where towing long streamers is not possible, a method called a *seafloor seismic survey* is typically used. In this arrangement, one or more hydrophone streamers are laid out in fixed positions on the ocean floor and the seismic vessel with its sound source moves above them.

Seismic imaging. Advanced computer technology has enabled three-dimensional (3D) seismic imaging. This approach may increase the likelihood of successfully locating a reservoir by as much as 50%. The seismic imagery can be presented to one or more analysts (sometimes wearing special goggles) in a true 3D environment (fig. 4–8). Images can be moved and rotated as desired, and the analysts in some cases can actually walk around inside the image itself.

Fig. 4–7. Seismic survey vessel (*Source:* Leffler, Pattarozzi, and Sterling)

Fig. 4–8. Reservoir location by 3D seismic imaging (*Source:* Statoil)

In addition, four-dimensional (4D, or time-lapse) imaging can show the movement over time of oil and other subsurface fluids, as well as the progress of drilling or logging tools.

The analytical and survey techniques described above merely provide indications of where oil or gas accumulations might be found. The only way to prove the actual existence of oil or gas in the subsurface is to drill an exploration well and test the contents of the target formation.

After an E&P company has identified a formation that shows promise, it must then determine more precisely the extent and orientation of the reservoir, the composition and related properties of the crude oil, and the properties of the geologic formation that could affect later (developmental) drilling and production. The shorthand term for this process is formation evaluation, and it typically begins with exploratory drilling.

Exploratory Drilling

The company generally begins a formation evaluation program by obtaining required leases from landowners or other parties and then drilling one or more exploratory wells. (An exploratory well drilled far from any known hydrocarbon deposits is called a *wildcat*.) Exploratory drilling is undertaken to locate probable mineral deposits or to establish the nature of geologic structures. Such wells may not be capable of production if minerals are discovered.

(Chapter 6 discusses development drilling. In contrast to exploratory drilling, development drilling is conducted to delineate the boundaries of a known mineral deposit to increase the productive capacity of the formation.)

Selecting the number and location of exploratory wells is critical, because drilling is very expensive. Onshore wells can cost as much as $15 million depending on well depth, while a 100-day offshore drilling program in deep water can cost $100 million.

The process can also be time-consuming. An API study found that it could take as few as 3 years or as many as 10 years to explore, develop, and begin production from an oil or gas lease on federal lands in the U.S. Intermountain West.[2]

DOE tracks the number of wells drilled each year in the United States. Data for recent years are shown in figure 4–9 and table 4–1.

It is difficult to generalize about the success rate of exploratory drilling because of wide variety in the type of formations and locations where such projects are undertaken. However, API has stated that the exploration drilling success rate nearly tripled from 1990 to 2007 (from about 26% to 71%) owing to advances in technology.[3] DOE reported a success rate of 65.1% in 2008.[4]

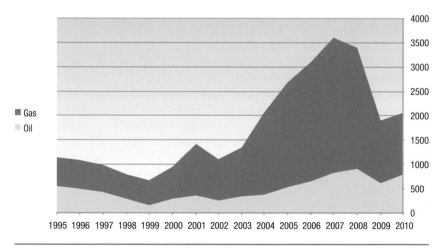

Fig. 4–9. U.S. exploratory wells drilled (oil and gas): 1995–2000 (*Source:* EIA)

Table 4–1. U.S. drilling activity: 1997–2010

	Year	Crude oil	Gas	Dry holes	Total	Success rate,[a] %
Exploratory	1997	491	562	2,113	3,116	33.3
	2000	287	657	1,340	2,284	41.3
	2002	258	845	1,281	2,384	46.3
	2005	540	2,134	1,467	4,141	64.6
	2008	921	2,467	1,787	5,175	65.1
	2009	626	1,290	1,159	3,075	62.3
	2010	938	1,269	1,364	3,571	61.8
Development	1997	10,715	10,935	3,761	25,411	86.4
	2000	7,804	16,383	2,800	26,987	89.6
	2002	6,514	16,487	2,456	25,457	90.3
	2005	10,229	26,482	3,190	39,901	92.0
	2008	15,908	30,959	3,680	50,547	92.7
	2009	11,057	17,712	2,657	31,426	91.5
	2010	18,912	17,973	3,321	40,206	91.7
Overall	1997	11,206	11,497	5,874	28,577	79.4
	2000	8,091	17,040	4,140	29,271	85.9
	2002	6,772	17,332	3,737	27,841	86.6
	2005	10,769	28,616	4,657	44,042	89.4
	2008	16,829	33,426	5,467	55,722	90.2
	2009	11,683	19,002	3,816	34,501	88.9
	2010	19,850	19,242	4,685	43,777	89.3

Source: EIA, *Monthly Energy Report,* February 28, 2011, table 5.2.
[a] Success rate = (total wells drilled − no. of dry holes) ÷ total wells drilled.

A key objective of exploratory drilling is to determine the depth and thickness of pay zones—vertical sections of a geologic formation whose pores may hold oil and gas (and probably some water) in varying concentrations. Descriptions of the drilling systems used for exploratory drilling (as well as for development drilling) are presented in the next two chapters (5 and 6).

Evaluation Methods and Tools

During the drilling of an exploratory well, a variety of measurement and assessment operations are carried out. These include the use of logging, real-time methods, and other tools.

Well logging

Logging refers to the performance of tests during (and sometimes after) the drilling process that allow geologists and drill operators to gain a clearer picture of subsurface formations and to monitor drilling progress. More than a hundred different logging tests can be performed.

Lithographic log. A lithographic log is a physical description of the rock through which the well is being drilled (fig. 4–10). The information for this kind of log comes primarily from well cuttings—small chips of rock created by the drill bit and brought up to the surface by a circulating fluid called *drilling mud*. Geologists take samples at regular intervals and study the chips under a microscope. They then prepare a written description of the well, showing what kinds of rock were found at what depth.

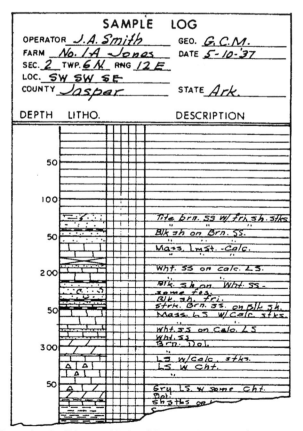

Fig. 4–10. Example of a lithographic log (*Source:* Hyne, N. J.)

A more precise lithographic log may be desired for a subsurface interval of interest. To prepare such a log, workers can use a special drill to extract a core—a cylinder of rock typically two to five inches in diameter and 20–90 feet long—and bring it to the surface intact for study.

In *rotary coring*, the bottom end of the drill (fig. 4–11) has a bit comprising sharp teeth of tungsten steel or industrial diamonds arranged around its outer edge. It cuts in a circle around the outside of the wellbore. As the drill moves downward, the rock core passes up through the hollow bit and into a cylinder called a *core barrel*, mounted right above the bit.

Fig. 4–11. Rotary coring drill bit (*Source:* Hyne, N. J.)

Alternatives to rotary coring punch or drill holes outward into the rock formation, perpendicular to the wellbore. One such approach is *percussion sidewall coring*, in which multiple small *core tubes*, or *bullets*, are fired into the sidewall to take samples. The tubes are then pulled back into the coring tool by wires and brought back up to the surface for study.

Another alternative is *rotary sidewall drilling*. As the name implies, a small rotating bit drills into the sidewall to obtain a sample. The drill bit can be moved to take multiple samples, each of which is captured and segregated within the tool prior to retrieval.

Normal drilling must be suspended to conduct a coring operation, so the process is expensive. In addition, the deeper the well, the more expensive it is to obtain a core sample.

Drill–time log. A *drill-time log*, as the name implies, records the rate of penetration (ROP) of the drill as it moves down through the rock layers. Drilling time is a function of both drilling parameters (e.g., drill rotation, weight on the drill bit, and type of bit) and rock properties. Assuming drilling parameters are held constant, changes in ROP can indicate a change in the type of rock encountered by the bit.

Mud log. A mud log records the chemical analysis of drilling mud and well cuttings as a well is being drilled, looking for traces of hydrocarbons. Correlating this analysis with drilling depth can help to pinpoint rock formations that bear gas or oil.

Wireline logs. Several kinds of analyses can be conducted with wireline logs. Instruments of various kinds are mounted inside a special cylinder called a *sonde*—typically 25–75 feet long and 3–10 inches in diameter— that is moved slowly up and down a well on a flexible electrical cable, or *wireline*. In preparation for wireline logging, the drill string is removed, and the wellbore is typically filled with drilling mud.

Electrical, radiation-based, acoustic, and ultrasonic instruments record such parameters as resistivity, rock type, porosity (and pore size), density, fluids content, and even the orientation of layers in the rock (table 4–2). Signals from the sonde are transmitted through the wireline to the surface, where they are recorded. From these data, logs are created that shows the continuously recorded measurements, at all points, throughout the depth of the well. The logs help to determine the extent and continuity of reservoirs and to assess the magnitude of oil and gas reserves.

Real-time methods

As noted above, wireline logging is done after a well has been drilled. However, companies can also now take real-time measurements and conduct logging while drilling.

Table 4–2. Wireline tool types

Measurement type	Measurement mechanism	Parameters assessed
Gamma	Natural gamma radiation	Lithology (correlation); presence of potential reservoir rock; shale content
Density	Bulk density	Porosity; lithology
Neutron	Hydrogen index	Lithology; porosity; gas indicator
Acoustic	Sound travel time; acoustic waveform	Porosity; lithology; seismic calibration
Resistivity	Electrical resistance of formation	Saturation; permeability
Induction	Induced electrical current level	Saturation
Image	Resistivity or acoustic pixelated image	Sedimentology; fracture/fault analysis
NMR	Nuclear magnetic resonance	Porosity; pore size; permeability; saturation
Formation tester	Pore pressure	Fluid types; pressures and contacts
Spontaneous potential	Electrical current between fluids of different salinity	Presence of potential reservoir rock
Caliper	Wellbore dimension	Wellbore diameter; volume

Source: Adapted from Hyne, N. J.; and Jahn, Parrish, and McCartney

Measuring while drilling (MWD). Properties such as azimuth and deviation of the wellbore can be tracked by MWD systems. They can create a directional log, using magnetometers to show the orientation of the drill bit and the direction in which the well is being drilled.

Logging while drilling (LWD). LWD tools can measure virtually all of the parameters described above in the section on wireline logging. Sensors are located just above the drill bit, powered by batteries or by small turbines driven by circulating drilling mud. Data can be sent to the surface as a series of digitally coded pressure pulses, where they are decoded by a computer.

Other evaluation methods

In addition to the search, survey, and logging methods described above, companies also use other procedures to assess the characteristics and production potential of hydrocarbon-bearing formations.

Cuttings evaluation. As noted above in the section on lithographic logs, the analysis of rock cuttings (also called well cuttings) are an important evaluation tool. During well drilling, rock cuttings are usually collected every one to five feet. These cuttings are small fragments of rock that provide clues as to the presence of hydrocarbons; they are examined closely for evidence ("shows") of oil and gas.

Cuttings and cores also are useful in evaluating the conditions under which the rock layers were first formed and later transformed to provide favorable conditions for hydrocarbon generation, accumulation, and trapping.

Drill stem flow testing. Drill stem flow testing is the direct sampling of a portion of a well to check for the presence of oil or gas. A vertical section (pay zone) of the well is temporarily *completed* (sealed above and below the zone to be tested); drilling mud is removed; and water, gas, and oil are allowed to flow into the section. A valve on the drill stem tool allows these fluids to flow into the tool. Any gas present will generally flow up through the tool (and the drill string above it) to the surface for measurement. Oil may do the same if formation pressure is sufficient, or it may just fill the drill stem to a certain level that is measured later.

Pressure within the drill stem tool is continuously monitored. The tool's valve is opened and closed several times, and a record of pressure change is made for a period of time that can range from a few minutes to a few days. The pressure record can help determine permeability, reservoir fluid pressure, and the presence of any damage to the formation.

After drill stem testing ends, drilling can resume.

In summary, data gathered in the exploratory drilling phase and the subsequent formation evaluation process are used to decide whether to go ahead with a development (production-oriented) drilling program. A company must determine whether the oil or gas resources it has found are sufficiently large and of sufficient quality that they can be profitably produced.

Making the Decision to Begin Production

A variety of tools have been created for analysis of the economics of hydrocarbon development and the assessment of related risks. As noted, development of an oil or gas project is, by definition, a high-cost venture with many significant uncertainties.

Although it is beyond the scope of this book to describe these tools in detail, the following sections provide a broad picture of the issues that they address. Note that no attention is given to the economics and risk of "start-from-scratch" exploratory activity—which, despite major financial investment, can result in zero return if no commercially producible oil or gas accumulations are found.

Development economics

At the outset of a project, discussion with key stakeholders can be vital to identify opportunities for collaboration, potential issues of concern, and procedures for obtaining needed approvals. Stakeholders may include government and regulatory bodies, development partners, neighboring property holders, nongovernmental organizations (e.g., environmental or human rights groups), financial institutions, and suppliers and contractors.

On the financial side of the ledger, estimates must be developed for:

- Capital costs (production platforms; drilling gear; and production facilities, e.g., compressors, pumps, separators, and instrumentation)
- Operating costs (including for maintenance and workovers)
- Personnel needs (workforce size and skills, labor costs, and overhead)
- Expected revenues (from oil and gas sales, tariffs, and payments from partners)

Special attention must be paid to agreements with a host government in regard to financial issues, such as taxes, royalty rates, royalty payment method, financial liabilities, and involvement in social development or training programs. Related issues include expectations for inflation, future oil and gas prices, and exchange rates.

Technical and operational factors

A company also must consider a range of issues related to the extraction and subsequent processing of oil and gas. These include:

- Expected rate of decline in well production over time
- Expected oil and gas quality (and required cleanup/treatment actions)
- Environmental regulations
- Proximity of storage facilities and pipelines (or other transportation methods) for getting oil and gas to market
- Planning for possible operational problems during drilling and production
- Decommissioning of equipment and sites at the conclusion of the project.

References

1 American Petroleum Institute. 2011. *The Story of Oil and Natural Gas.* Interactive Web presentation. http://www.api.org/story/ (accessed April 2, 2011). Washington, DC: American Petroleum Institute.

2 American Petroleum Institute. 2011. *Explaining Exploration and Production Timelines (Onshore).* Washington, DC: American Petroleum Institute.

3 American Petroleum Institute. 2008. America's energy savers. http://www.api.org/ehs/climate/new/upload/Energy_Efficiency.pdf (accessed April 2, 2011).

4 U.S. Department of Energy, Energy Information Administration. 2011. *Monthly Energy Report.* Washington, DC: U.S. Department of Energy.

5 Drilling and Completion: Onshore

Following the exploratory drilling and formation evaluation efforts described in the previous chapter—and assuming a positive result from assessment of risks and potential rewards—the E&P company then begins an expensive campaign of development drilling. The objective is to drill down to those formations that show promise and to put into place the equipment and systems needed to produce oil and gas safely and cost-effectively.

This chapter focuses on onshore (or land) drilling. Discussion of offshore drilling follows in chapter 6.

Site Preparation and Well Construction

Typically, each onshore well is sited within a drilling spacing unit (DSU). In recent historical practice, a DSU was a square encompassing 40 or 80 acres, within which only one well was drilled. This constraint was established in the 1930s in the United States to prevent exploitation of a field with excessive drilling and production rates.

More recently, based on advances in geologic evaluation and drilling technology, tighter spacing is being used, depending on such factors as formation permeability and heterogeneity, and drilling costs. A 10-acre DSU is not uncommon. In addition, operators can now drill as many as two dozen wells in a spoke pattern from a single surface location (pad).

Once the drilling site has been selected, several preparatory steps are taken. Workers establish the boundaries of the work area and carry out environmental impact studies if necessary. Lease agreements, titles, and right-of-way accesses also may be sought.

After all legal and environmental issues are settled, the crew goes about preparing the drill site. For an onshore site, this entails the following steps:

- The land is cleared and leveled, and access roads are built if needed.

- A water well is drilled if necessary to provide the significant volumes of water used in oil and gas drilling operations.

- A reserve pit is dug and lined with plastic, to hold rock cuttings and drilling mud generated during the drilling process. In environmentally sensitive areas, the cuttings and mud are trucked off-site instead of being placed in the pit.

As described in chapter 6, for an offshore site, the drill site is on the seabed. There is no site preparation in the conventional sense.

Once the land has been prepared, the crew digs several holes to make way for the rig and the main hole (fig. 5–1). A rectangular pit, called a *cellar*, is dug around the point where drilling will take place. The cellar provides a workspace around the hole and room for auxiliary equipment that will be located below the floor of the main drilling platform.

The crew then drills the main hole, often with a small drill truck rather than the main rig. The first part of the hole is larger in diameter than, and not as deep as, the main portion will be and is lined with a large-diameter conductor pipe. Other holes are dug off to the side to temporarily store equipment. After these holes are finished, the rig equipment can be trucked in and set up. If necessary, equipment may be brought to the site by helicopter or barge.

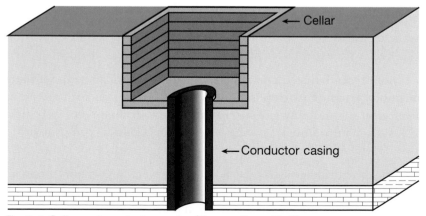

Fig. 5–1. Cellar and conductor casing (*Source:* Hyne, N. J.)

Some history: Cable tool drilling

The first commercial oil well drilled in the United States—in 1859 at Titusville, Pennsylvania—used a cable tool rig. All U.S. fields discovered during the 1800s were drilled by cable tool rigs, and as recently as 1953, New York Natural Gas Corporation used such a rig to drill a well to 11,145 feet in New York State.

In simple terms, a cable tool rig repetitively raises and drops a solid steel rod (about 5 feet long with a chisel point on its lower end) to pound, rather than drill, a hole into the ground (fig. 5–2). The bit is raised occasionally so that water and rock chips can be removed.

Progress using a cable tool rig was very slow—about 25 feet per day on average. Further, the inability to control subsurface pressure meant that blowouts were frequent.

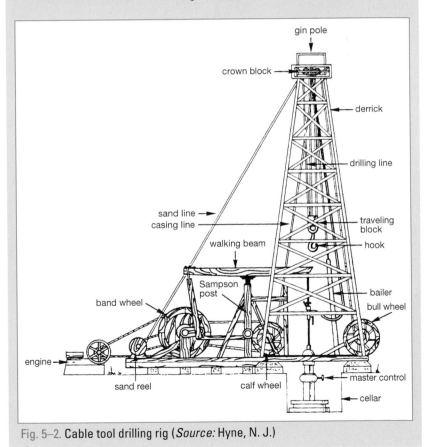

Fig. 5–2. Cable tool drilling rig (*Source:* Hyne, N. J.)

Rotary Drilling

Virtually all oil and gas wells—onshore and offshore—are drilled using a system called a rotary drilling rig, which can drill several hundred to several thousand feet per day. This system turns a long length of steel pipe with a sharp bit on its lower end to cut the wellbore (fig. 5–3). A basic rotary drilling system consists of four groups of components: prime mover (one or more engines), hoisting equipment, rotating equipment, and circulating equipment. Many innovations have been developed since rotary drilling first came into widespread use in the early 1900s.

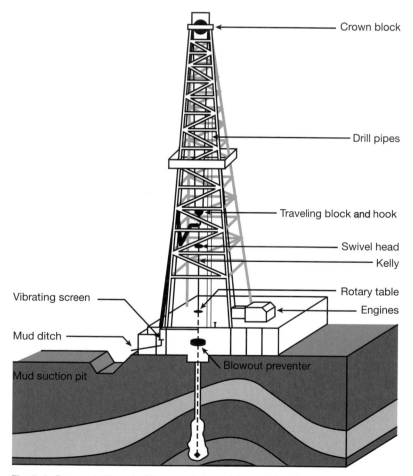

Fig. 5–3. Rotary drilling rig (*Source:* Adapted from Jahn, Cook, and Graham)

Prime mover

Most rotary rigs use from one to four diesel engines, generating as much as 3,000 horsepower. This prime mover powers the rotating equipment, the hoisting equipment, and the circulating equipment, as well as associated lighting, water-pumping, and compression equipment.

Hoisting equipment

The hoisting equipment (fig. 5–4) raises and lowers what goes into or comes out of the wellbore. The most visible part of the hoisting equipment is the derrick, or mast, a towerlike structure up to 200 feet tall. The floor of the derrick is typically built on a framework that raises it 10–30 feet above the ground, to allow space for installation of wellhead equipment.

A strong hoisting line (braided steel wire wound on a fiber or steel core) is spooled around a frame, called the *drawworks*, on the derrick floor. The prime mover drives the drawworks to wind and unwind the hoisting line.

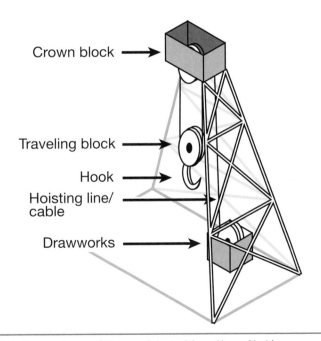

Crown block →

Traveling block →

Hook →

Hoisting line/
cable →

Drawworks →

Fig. 5–4. Hoisting equipment (*Source:* Adapted from Hyne, N. J.)

From the drawworks, the line goes up to the top of the derrick and attaches to a set of large pulleys (bolted to the derrick) in a block-and-tackle arrangement. Below the block (called the *traveling block*) is a large hook, from which a swivel hangs. As the drawworks reels in or pays out the hoisting line, the equipment hanging from the hook moves up or down.

Rotating equipment

Rotating equipment (fig. 5–5) includes almost every piece of hardware that goes into or out of the wellbore. A key component—hanging directly from the hook described above—is a very strong pipe called the *kelly*. Made of high-grade molybdenum steel, it is a standard length (40 or 54 feet) and has four or six flat sides. Those flat sides allow the kelly's lower end to fit into (but slide up and down within) a special fitting (the *kelly bushing*) that is attached to a circular device on the derrick floor called the *rotary table* (fig. 5–6).

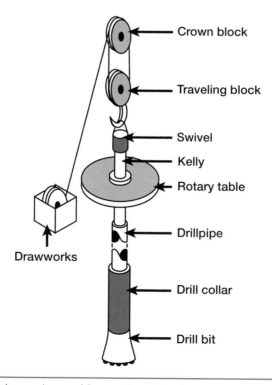

Fig. 5–5. Rotating equipment (*Source:* Hyne, N. J.)

Fig. 5–6. Kelly, kelly bushing, and rotary table (*Source:* Hyne, N. J.)

At the very start of a drilling operation, one section of round drill pipe is screwed tightly to the bottom end of the kelly. That first section—with a sharp drill bit attached to its lower end—is then lowered through the bushing and rotary table until the flat sides at the lower end of the kelly are seated securely in the kelly bushing.

At that point, the kelly, bushing, and rotary table begin to rotate as a single unit, powered by the prime mover. The drill bit (turning clockwise when viewed from above) begins to bite into the ground below the derrick.

Each drill pipe section is called a joint. Made of heat-treated alloy steel, each section can be from 8 to 45 feet long (commonly 30 feet), with an outside diameter ranging from about 3 to 5.5 inches. Collectively, the downhole drill pipe, the bit, and related equipment comprise the drill string (fig. 5–7).

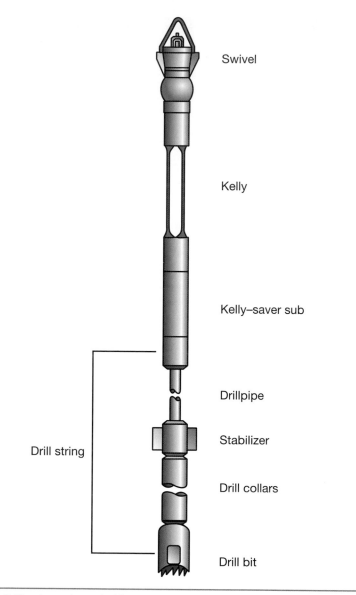

Swivel

Kelly

Kelly–saver sub

Drillpipe

Drill string

Stabilizer

Drill collars

Drill bit

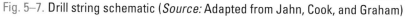

Fig. 5–7. Drill string schematic (*Source:* Adapted from Jahn, Cook, and Graham)

As the drill bit goes deeper, additional sections of drill pipe must be added. Drilling stops temporarily while workers carefully choreograph the following actions:

- The drill string is raised until the kelly is completely above the rotary table.

- The topmost drill pipe section ("A") is grasped with special tools so it cannot fall back into the wellbore, and the kelly is unthreaded from section A.
- A new section ("B") is swung into place and its top is screwed into the kelly.
- The bottom of section B is screwed into the top of section A.
- The kelly is lowered to reengage with the kelly bushing, at which point drilling resumes.

Note that special sections can be attached to the drill string, as needed, to perform desired functions, including the following:

- Connecting the drill bit to the first section of drill pipe
- Adding weight to the drill string (to increase drill bit pressure on the rock)
- Absorbing vibration or shocks transmitted from the drill bit up the drill string
- Stabilizing the drill string within the wellbore
- Connecting drill pipe sections that have different diameter or different threads

Automated systems for handling drill strings (as well as casing, described below) are commercially available to help improve safety as well as efficiency in drilling operations.

Drill bit

The business end of a drill string is the drill bit, and different rock layers often require the use of different drill bits to achieve maximum drilling efficiency. Several designs have evolved as companies gained experience drilling in various kinds of geologic formations. Drill bits commonly range in diameter from about 4 to 26 inches.

The most common type is the rotary cone bit, and the general design that uses three rotating cones is called a *tricone* bit. As the drill string turns, the three cones also rotate, and teeth or buttons on the cones either flake or crush rock at the bottom of the well. Typical rotation speeds range from 50 to 100 rpm.

There are hundreds of different tricone bit designs, generally classified as either *milled-teeth* (fig. 5–8) or *insert* (fig. 5–9) bits. The former are most suitable for rock of soft or medium hardness; the latter are more effective on hard rock.

Fig. 5–8. Tricone milled-teeth drill bit (Source: Hyne, N. J.)

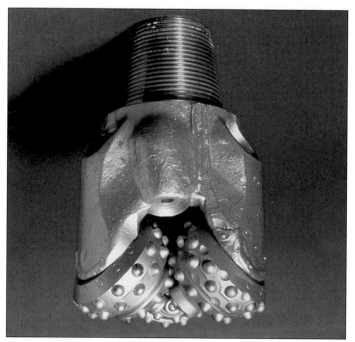

Fig. 5–9. Tricone insert drill bit (*Source:* Hyne, N. J.)

A different drill bit design uses industrial diamonds to shear away the rock. Its metal cutters are studded with chips of diamond (rather than milled teeth or tungsten carbide insert buttons). The diamond drill bit has no moving parts (unlike the rotating cones in the tricone design). It simply turns at the same speed as the drill pipe to which it is attached.

A common configuration is the polycrystalline diamond compact (PDC) drill bit (fig. 5–10). Although a PDC bit is expensive, it can last for several hundred hours and drill more footage than other types.

The structures that connect a drill bit to the drill string above it contain channels. These enable the flow of drilling mud to lubricate and cool the drill bit and flush rock cuttings from the face of the bit.

The weight of the drill string pressing down on the drill bit is controlled from above by the drawworks and the hoisting line, as well as by the number of sections added to the drill string. Typically, 3,000–10,000 pounds per square inch (psi) of pressure per inch of bit diameter is applied.

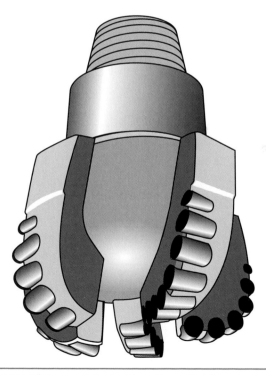

Fig. 5–10. **PDC drill bit** (*Source:* Adapted from Hyne, N. J.)

After extended use, a drill bit must be changed because of wear or breakage. A tricone bit, for example, is commonly changed after 24–48 hours. In other cases, a different drill bit may be needed to penetrate a different type of rock.

In a time-consuming process called *tripping*, drilling is halted while the entire drill string is pulled from the wellbore. The drill pipe sections are unscrewed in sequence (sometimes three sections at a time) and are stored temporarily on a special rack on the drilling platform. A device called a *bit breaker* is attached to the rotary table to grip and unscrew the bit when it reaches the surface. A new drill bit is then screwed into place; the drill string is reassembled in order and is lowered back into the hole; and drilling resumes.

As a drill bit moves downward, it encounters increasing temperature and pressure. On average, temperature rises by about 1–2°F for every 100 feet of depth below the surface. For example, at 10,000 feet, oil in a reservoir below the Anadarko basin, in Oklahoma, would be at about 190°F. Pressure increases at a rate of about 45 psi per 100 feet of depth. Therefore, at 10,000 feet, a pressure of about 4,500 psi would be expected.

As noted below (in the "Drilling techniques" section), there are situations in which the drill bit can be made to rotate independently from the drill string, driven by a specially designed downhole motor.

Circulating system

The circulating system consists of a range of equipment that works together to move drilling fluid (called *mud*) into and out of the hole being drilled. This drilling mud serves several purposes: it cools and lubricates the drill bit; it flushes cuttings and debris away from the face of the drill bit; and it coats the walls of the well to stabilize them.

Another critical function of the circulating system is to help control pressure within the well. The weight of the drilling mud can be adjusted to exert greater pressure at the bottom of the well than the pressure exerted by fluids (oil, gas, and water) in the surrounding rock. This control is necessary to prevent uncontrolled fluid flow into the well and to prevent the well walls from caving in, trapping equipment downhole.

Drilling mud can consist of any of the following mixtures:
- Water (fresh or saline) and clay (commonly bentonite)
- Oil (diesel, mineral, or synthetic) and clay

- Water with 10% oil, plus clay (emulsion)
- A synthetic organic material and water

Chemicals and other additives to the drilling mud are used to adjust such parameters as weight (density), viscosity, and pH.

Drilling mud is mixed and stored in large steel tanks near the drilling rig. Rotating paddles or a high-pressure jet constantly agitate the mud to ensure uniform consistency. Large pumps (driven by the prime mover) move the mud through a long rubber hose to the swivel (above the top of the kelly) and then down into the hollow, rotating drill string. The mud sprays from nozzles or jets near the face of the drill bit.

Then, still under pressure from the mud pumps, the mud begins its trip back to the surface, flowing upward within the space between the rotating drill string and the walls of the well (called the *annulus*). At the top of the well, the mud moves into the mud return line and then through a series of devices that remove well cuttings, sand, silt, and even dissolved gases. This "clean" mud goes back into the mud tanks, ready for recirculation into the well, while removed solids are directed into a reserve pit.

After a well has been drilled, the mud cannot be reused. If made with freshwater, the mud can be spread on nearby land to fertilize crops. Other types must be moved by truck or barge to an approved disposal site.

Blowout prevention

A blowout is the uncontrolled flow of oil or gas up the drill string or wellbore annulus. This is triggered when the formation pressure (exerted by gas or oil in the formation around or at the bottom of the well) exceeds the pressure created by the drilling mud. A blowout can seriously injure workers and damage the drilling rig.

As noted above, the drill string is situated in the center of the wellbore, and there is a space (the annulus), typically several inches across, between the outer surface of the drill string and the wall of the well. For prevention of a blowout, it is important to at least cap the annulus. It may also be desirable to seal off the top of the drill string; however, in some cases, the operator may want to pump heavier mud down the drill string until mud pressure overcomes the pressure exerted by downhole formation fluids. For one or more of these goals to be achieved, a blowout preventer (BOP) is bolted securely to the top of the well, below the floor of the derrick (fig. 5–11).

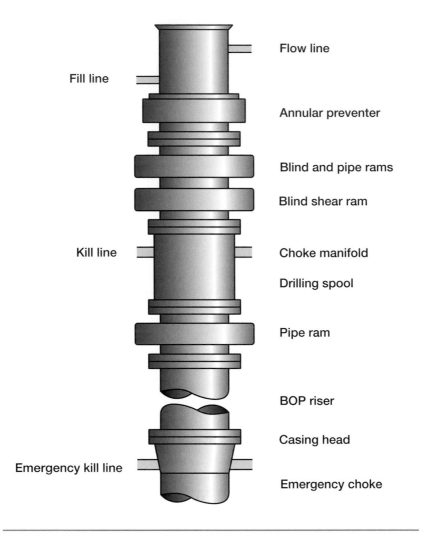

Flow line

Fill line

Annular preventer

Blind and pipe rams

Blind shear ram

Kill line

Choke manifold

Drilling spool

Pipe ram

BOP riser

Casing head

Emergency kill line

Emergency choke

Fig. 5–11. BOP (*Source:* Adapted from Jahn, Cook, and Graham)

The BOP—designed and built to rigorous standards set by API—typically consists of several stacked components. At the top is an *annular preventer*—a rubber-and-steel doughnut that can be rapidly compressed hydraulically to close around equipment of any size or shape still in the well. The aim is to completely seal the annulus.

If the annular preventer cannot stop the flow—and if there is no drill string or other equipment in the wellbore—then a series of large steel plates are rapidly pushed together to completely seal off the top of the well. The first of these is called a *blind ram* and is mounted immediately below the annular preventer.

If the blind ram is ineffective (or cannot be used), then a *pipe ram* (mounted below the blind ram) is activated. The pipe ram also is composed of two large blocks of steel, but each has a semicircular cutout in its mating surfaces. These cutouts allow the ram to close around, but not damage, the drill string (in the same manner as the annular preventer) while simultaneously sealing the annulus. Recall that the operator may want to pump mud down the drill string to help control downhole pressure.

A *blind shear ram* consists of two sliding steel plates with intersecting blades. As the plates come together, the angled blades cut the drill string; then, the plates in which they are mounted move into place to seal both the drill string and the annulus. Specialists who studied the April 2010 blowout of BP's deepwater Macondo well in the Gulf of Mexico reported in March 2011 that one key part of the well's 300-ton, 50-foot-tall BOP, about 4,000 feet below the water's surface, failed to operate properly. In particular, Det Norske Veritas reported that a piece of drill pipe had buckled and became wedged in the blind shear ram as it tried to close. The pipe had been pushed out of its normal position by the flow of oil and gas up the well. Because the ram could not cut the pipe and fully close, the well was not sealed off.

All BOP components can be activated separately by hydraulic pressure, maintained in accumulator cylinders. Their operation does not depend on the rig's prime mover. Workers can trigger the BOP from any of several control stations on or near the drilling rig.

A final BOP-related component is the *choke manifold*, a set of flow lines, valves, and chokes connected to the BOP stack. It is routinely used for nonemergency routing of drilling mud from the well to various onsite equipment. However, if the BOP has been activated, it can also be used to relieve well pressure and to circulate heavier drilling mud into the well.

Drilling techniques

Every well operator would like to drill a straight well directly down into the earth, expending the least amount of time and money to reach a subsurface target. In reality, because of the physics of rotary drilling, a straight hole in relatively uniform ground typically has a slight corkscrew configuration, with the drill bit moving downward within a cone whose angle is as large as five degrees. In most cases, this digression from true vertical is acceptable.

Directional or deviated drilling. In many situations, however, it may be necessary or desirable to drill at a predetermined angle to avoid (or reach) specific rock formations or orientations or to hit a desired target. This is called *directional* or *deviated* drilling (fig. 5–12). This deviation from true vertical can be initiated at the start of drilling or after a vertical well (or well segment) has already been drilled. In the latter case, special tools are used—such as a long steel wedge (called a *whipstock*), or a device that blasts drilling mud at high pressure at the desired angle against the wall of the wellbore—to redirect the path of the drill bit.

A more advanced method for directional drilling sends a stream of high-pressure drilling mud down through a pipe to power a specially designed downhole *mud motor* that has a drill bit attached (fig. 5–13).

The pipe does not spin, only the drill bit, as a pilot hole is drilled at the desired orientation.

After the new angle is confirmed as correct, the original mud-motor system is replaced by a more elaborate unit called a downhole steerable assembly (fig. 5–14). This assembly can maintain or further change the deviation angle.

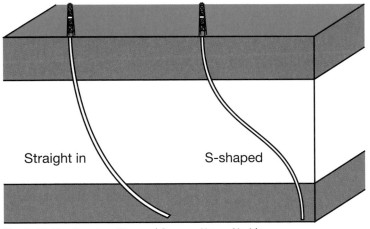

Fig. 5–12. Deviated wellbores (*Source:* Hyne, N. J.)

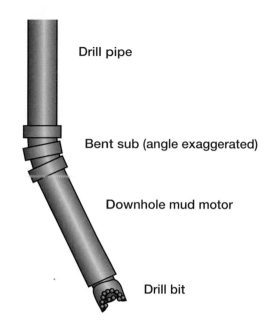

Drill pipe

Bent sub (angle exaggerated)

Downhole mud motor

Drill bit

Fig. 5–13. Downhole assembly to start drilling a deviated wellbore (*Source:* Adapted from Hyne, N. J.)

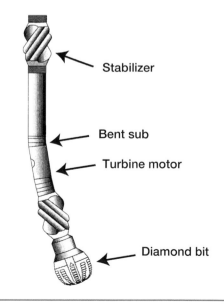

Stabilizer

Bent sub

Turbine motor

Diamond bit

Fig. 5–14. Steerable downhole assembly for deviated drilling (*Source:* Hyne, N. J.)
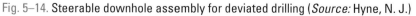

Deviated drilling is particularly valuable in offshore operations because it is very expensive to build and operate multiple drilling platforms and wells to develop a large offshore field (particularly in deep water). Instead, a large production platform is typically constructed to drill several main wells, with many deviated or directional wells drilled out from each of those wells, at various depths and angles.

One notable example is the *Cognac* platform, sited in 1978 by Shell Oil in the Gulf of Mexico. Workers on the Cognac platform put in place 62 wells, all of which used deviated drilling. (The platform is still in operation as of early 2012.)

Advances in drilling technology

In the United States, Gas Research Institute (now Gas Technology Institute [GTI]) partnered with the DOE, oil and gas E&P companies and analysts as far back as the mid-1980s to develop geologic information as well as technology and procedures for both finding and evaluating the potential of natural gas chiefly from unconventional sources, such as coal seams, tight (low–permeability) sands, and shales.[1-3]

Collaborative GTI programs also were instrumental in spurring the development of new technologies for tapping and optimizing production from unconventional gas resources. These programs included:

- Documentation of successful drilling practices and compilation of environmental impact assessment tools for operators in a dozen U.S. basins/formations
- Advancement of technologies for slimhole, microhole, coiled-tube, horizontal, directional, jet-assisted, laser, and underbalanced drilling
- Evaluation of new drill bit designs, hydraulic fracturing methods, wellbore perforation technology, novel cement compositions, and downhole (cross-well) seismic imaging

Precision drilling

Shell Oil (the U.S. subsidiary of Royal Dutch Shell) brought into operation its *Perdido* production platform in March 2010, located nearly 200 miles offshore in the Gulf of Mexico, in water almost two miles deep.[4] The company reported that operators on the *Noble Clyde Boudreaux* floating drilling rig, using advanced control systems, were able to guide the drill bits as much as two miles farther below the seabed to hit targets the size of the lid on a garbage can.

Extended-reach (ER) drilling. A specific type of deviated drilling is ER drilling. An ER well has just a single deviation from vertical and typically bottoms out thousands of feet horizontally from its point of origin (fig. 5–15). This technique can be particularly useful in drilling from an onshore site to a target lying below the seabed offshore or in tapping resources that lie beneath environmentally sensitive areas.

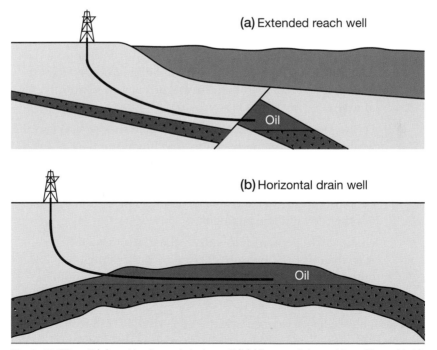

Fig. 5–15. ER well (a) and horizontal-drain well (b) (*Source:* Adapted from Hyne, N. J.)

ExxonMobil and its international partners announced in February 2011 that they had drilled the world's longest well, Odoptu OP-11, in Russia. This ER well set world records both for total measured depth (40,502 feet [7.67 miles]) and for horizontal reach (37,648 feet [7.13 miles]). Odoptu is one of three large fields five to eight miles offshore northeast Sakhalin Island, which lies off Russia's Pacific Coast.

Horizontal-drain well. A horizontal-drain well is a deviated well with a long horizontal section that taps into a subsurface structure whose hydrocarbon-bearing zones are parallel to the surface of the ground (see fig. 5–15). This approach is particularly effective in reservoirs with vertical fractures and in tight (low-permeability) formations.

Laterals. Short horizontal branches drilled outward from a main vertical wellbore are called laterals. They allow operators to tap hydrocarbon-bearing zones at different depths.

Drilling problems

A variety of things can go wrong while drilling a well. Examples include the following:

- A piece of equipment breaks off or falls into the wellbore.
- The drill string gets stuck in the well.
- Circulation of drilling mud is lost as the mud flows outward into a very porous, highly fractured, or cavernous layer (zone) of rock.
- Drilling mud is forced into any permeable rock adjacent to the wellbore, degrading permeability and hindering flow of gas or oil into the wellbore.
- Corrosive gases (e.g., carbon dioxide or hydrogen sulfide) flow out of the formation and attack the integrity of the drill string or other downhole equipment.
- Unusually high pressure is encountered in a formation, overwhelming the pressure exerted by the drilling mud. This causes a *kick*—a rapid, unexpected flow of formation fluids into the wellbore—which could lead to a blowout.

The industry has developed a wide range of tools and techniques to resolve these and related problems. However, their description is beyond the scope of this book.

Well Testing

A variety of well tests were described in chapter 4 as part of the exploration process, but other tests are also generally conducted during early production to help operators understand the nature of the well and its likely performance. In general terms, these tests help to determine the rate at which the greatest amount of oil or gas can be extracted without harming the formation.

Pressure is a particularly important parameter. It is measured in the tubing, within the casing, and at the bottom of the well. It is often measured while the well is flowing, but it is sometimes measured after a well has been *shut in* (i.e., flow has been deliberately stopped) for a day or so.

The operator will typically make an initial estimate of the well's *decline rate*. This parameter is an indicator of the expected drop in production rate as more and more oil or gas is extracted from a formation. Understanding the decline rate will guide decisions about the possible use of artificial lift to improve production (as described in chap. 7).

Other tests are run to determine:
- Potential (the maximum amount of oil and gas a well can produce in a 24-hour period)
- Productivity (various production rates and their likely effect on well life)
- Drawdown profile (change in bottom-hole pressure as a shut-in well is brought on production)
- Buildup profile (change in bottom-hole pressure as a flowing well is shut in)
- Absolute open flow (maximum flow rate of oil or gas into the well when the bottom-hole pressure is zero)

If a decision is made to recomplete a well (i.e., to tap a new zone above or below the original), a cased-hole log is commonly run to confirm the presence of oil or gas in the new zone. An instrument using a gamma-ray and neutron detector is typically used to assess the relative fluid content of the formation rock.

Various kinds of production logs are run to evaluate problems in flowing wells. Sophisticated instruments can detect fluid movement, measure fluid flow at various depths, listen for noises that differentiate liquid from gas, measure fluid temperatures, measure pressure at specific depths, measure water concentration in well fluids, and identify the best places to perforate well casing.

Onshore Well Completion

After an exploratory well has been drilled and tested—and the presence of commercially producible hydrocarbons has been verified—the well must be completed, in preparation for drilling a development well. The completion process is sometimes called *setting pipe*.

Often, a completion plan is prepared. This outlines the approach to be taken and may specify logging needs and casing and tubing dimensions (fig. 5–16).

COMPLETION PROGNOSIS

Well name: <u>Bobcat Wildcat # 1</u>   **Location:** <u>Mississippi</u>
<u>Canyon</u> <u>Blk 999</u>

Water depth: <u>7,500 ft.</u>

Well depth: <u>25,750 ft.</u>   **Kelly Bushing:** <u>125 ft.</u>

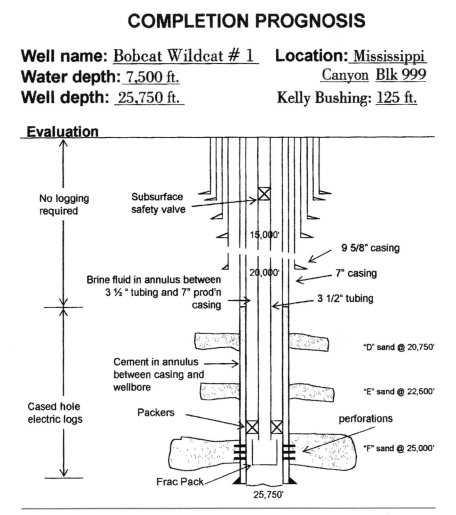

Fig. 5–16. Completion worksheet (*Source:* Leffler, Pattarozzi, and Sterling)

Casing

The first step in the completion process is to install casing—hollow steel pipe with relatively thin walls—from the surface to the bottom of the well. Casing can range from about 4 to 36 inches in diameter. Its outer diameter is at least 2 inches less than the diameter of the wellbore (which is generally wider at the top than at the bottom).

Sections of casing (typically 30 feet in length) are screwed together to form a long casing string (similar to the fabrication of the drill string), which is run into the wellbore and then cemented to the sides of the well (fig. 5–17). Casing serves several purposes: it prevents the sides of the well from caving in; it prevents flow of oil, gas, and salty water (produced while drilling) into underground freshwater reservoirs; and it prevents the flow of unwanted water and other fluids from the drilled formation into the gas and oil being produced.

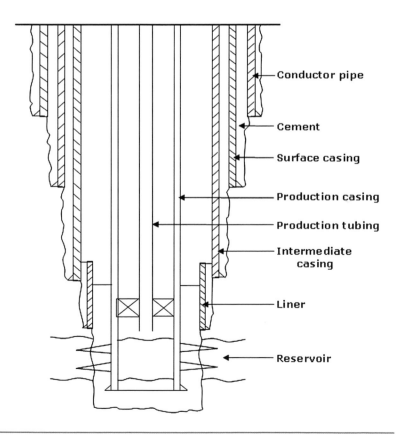

Fig. 5–17. Casing schematic (*Source:* Adapted from *Oil & Gas Journal*)

Hangers suspend the casing string from the top of the well (so that it does not go all the way to the bottom of the well), and downhole devices called *centralizers* keep the string centered within the wellbore. Once the casing string is in position, a slurry of cement (Portland cement with various additives) is mixed and pumped into the casing.

A device called a *cementing plug* or *wiper plug* is put into place before the cement is pumped. As the cement is pushed down into the casing, the plug moves ahead of it, wiping any material from the inner casing wall that might contaminate the cement slurry.

Movement of the cement displaces drilling mud downward. Equipment positioned at the bottom of the casing string prevents drilling mud from moving back up into the casing but allows the cement to flow out of the casing and then *upward* to fill the space between the casing and the walls of the well. The cement job is complete once the cement has hardened, after which the special tools are removed.

A related remedial operation, called a *squeeze job*, is sometimes conducted to force additional cement into place around the casing. It is used to seal leakage paths detected after initial cementing, to isolate a segment of casing that has been perforated (see the "Bottom-hole treatment" section, below), or to repair damaged casing.

In practice, a well is drilled, cased, and cemented in stages as the drill bit moves down into the earth. A well commonly has at least three concentric casing strings, though deeper wells have more. Each string runs all the way back up to the surface, with the largest-diameter (and shortest-length) string on the outside, and the smallest-diameter (and longest-length) string on the inside.

The largest-diameter casing string (30–42 inches)—also called *conductor pipe*—is typically several hundred feet long. It helps route drilling mud coming from the well back to the mud tanks, prevents the top of the well from caving in, and isolates any freshwater or gas zones that are near the surface. Proceeding down the well, the operator installs surface casing, intermediate casing, and finally production casing (each smaller in diameter than the casing above it).

Bottom-hole treatment

The very bottom of the well requires particular treatment to ensure maximum production of hydrocarbons. In an *open-hole completion*, the well is first drilled to the top of the hydrocarbon-bearing formation and casing is set. Then, drilling continues down into the formation, which leaves the bottom of the well uncased, or open.

In softer formations, where the walls of the well might cave into the well, a *gravel-pack completion* can be used, placing coarse sand (either loose or held within a special liner) at the bottom of the hole. This porous sand allows oil or gas, but not sediments, to flow into the well.

A common modern approach is the *set-through completion*. In this method, a liner or casing is cemented into the producing formation and is then perforated, using small, shaped explosives to create holes through the casing, the cement, and the formation. Oil and gas can then flow out of the formation and into the well.

Tubing

A final element of downhole equipment on a development well is tubing. This is small-diameter pipe (typically 1–4 inches), designed to carry gas, oil, water and other produced fluids to the surface. Sections of tubing (each about 30 feet long) are threaded together and positioned down the center of the well, inside the casing (fig. 5–18).

Fig. 5–18. Tubing being run into a well (*Source:* Hyne, N. J.)

The tubing performs several other important functions. For example, a rubber doughnut called a *packer*—installed at the bottom of the tubing—fits tightly around the tubing, large enough in diameter to tightly seal the space (annulus) between the outside wall of the tubing and the inside wall of the casing. Besides centering the tubing string, it prevents produced fluids from flowing up into the casing, where they could cause corrosion. Diesel oil or specially treated water is sometimes pumped into the annulus for added corrosion protection.

Because the tubing is suspended in the well (not cemented as the casing is), it can be pulled out relatively quickly for repair or replacement if needed. In contrast, removing and replacing casing is extremely expensive and time consuming.

Finally, a special narrow-diameter segment of tubing installed near the bottom of the string serves to stop any equipment that may fall into the tubing. If two or three *pay zones* (layers) in a subsurface formation are to be completed, the operator can install as many as three separate tubing strings, packers, and associated equipment to segregate the produced fluids from each zone.

Wellhead and related equipment

At the surface of the completed and cased well sits the wellhead—a large forged or cast steel fitting that is bolted or welded to the conductor pipe or surface casing. Equipment mounted above the wellhead called *casingheads* and *tubingheads* supports the downhole pipe strings and seals the annulus between the strings. Another typical unit is called a *Christmas tree*—an assembly of valves, fittings, pressure gauges, and chokes (fig. 5–19) designed to manage fluid flowing from the well.

Fig. 5–19. **Christmas tree at wellhead** (*Source:* Hyne, N. J.)

Well Treatment

Well treatment collectively refers to several techniques used to ensure the efficient flow of hydrocarbons out of a formation. These techniques are typically used to optimize production. However, they can also be used to repair formation damage—the blockage of pores in a formation by drilling mud, cement particles, corrosion products, or chemical buildup.

In broad terms, well treatment involves the injection of acid, gases, or water into the well to open up the rock formation and allow oil and gas to flow through it (and to the wellbore) more easily. Among the most common well treatments are acidizing and fracturing (also called fracing).

Acidizing

Acidizing is a process used to dissolve dolomite, limestone, or calcite cement between grains of sediment. Two types of acid are most commonly used—hydrochloric or hydrofluoric acid, sometimes mixed together—although acetic and formic acids are also used occasionally. Various additives are mixed with the acid to prevent both the corrosion of casing and tubing and the formation of iron compounds that could clog the pores of the reservoir.

Matrix acidizing is a well treatment used to enlarge the existing rock pores. In other cases, fracture acidizing is used, pumping the acid downhole under high pressure to both fracture and dissolve reservoir rock. Occasionally, an acid job is performed to clear away carbonate materials that can build up on the walls of a wellbore and inhibit oil and gas flow.

After an acid job, operators flush out spent acid and dissolved rock. The debris is pumped to the surface for treatment and disposal.

Hydraulic fracturing

As its name implies, hydraulic fracturing makes use of the force exerted by fluids (gases or liquids) under great pressure to crack open an underground rock formation, creating new fissures or expanding existing ones.

Background. Until the late 1940s, oil and gas well operators commonly used explosives to increase production, lowering nitroglycerin into a well and detonating it. The explosion fractured the rock, creating a large cavity. The cavity was then cleaned out, and the well was completed as an open hole.

According to API, the first commercial use of hydraulic fracturing was in 1947, and it has been used on more than one million wells since then. Since about 2006, increased use of fracing has been instrumental in boosting production from gas-bearing shale formations in North America. The technique is also finding wider use in the exploitation of oil-bearing shales. Fracing is often used in combination with horizontal drilling to reach as much of the formation volume as possible.

Energy in Depth, a group of U.S. oil and gas producers, has asserted that approximately 90% of the oil and gas wells in operation in 2011 had been fractured. And the National Petroleum Council estimates that 60%–80% of all oil and gas wells drilled in the United States from 2011 to 2020 will require fracture treatment to remain viable.

The fracturing process. The frac job is carried out under the direction of contractors who have specialized skills and equipment, as well as experience in performing such operations (fig. 5–20). A key ingredient is the frac fluid prepared for each job, brought in tanker trucks to the work site, where it is mixed and stored in large tanks. This fluid is typically water (though it can also be a gel, a foam, or a gas [e.g., nitrogen or carbon dioxide]), mixed with organic polymers and *proppant*—grains of quartz sand (sometimes resin coated), aluminum oxide pellets, or tiny ceramic beads. The injected fluid mixture is about 99% water and proppant.

When pumped into the ground under high pressure, the frac fluid pushes out into the rock formation, creating or widening existing cracks and crevices. The proppant props open the cracks after the pressure is reduced, enabling oil or gas held in the pores of the shale to flow more easily to the wellbore.

Predicting the exact geometry of a hydraulic fracture is complicated, so operators use a variety of tools to monitor fracture progress. In simple terms, the injected fluid is pressurized to a level greater than that imposed by the overlying rock. The term *fracture gradient* is typically used to quantify the required minimum pressure. A gradient of 0.8 psi per foot of depth means that that at 10,000 feet, it would take 8,000 psi of hydraulic pressure to create or extend a fracture.

Pumper trucks are used to create the desired downhole pressure. A small frac job can use 45,000 gallons of frac fluid and 35 tons of sand; an especially large job can require several million gallons of fluid and 1,500 tons of sand. Pumping pressure can reach 15,000 pounds per square inch, with fluid flow rates of 100 barrels per minute.

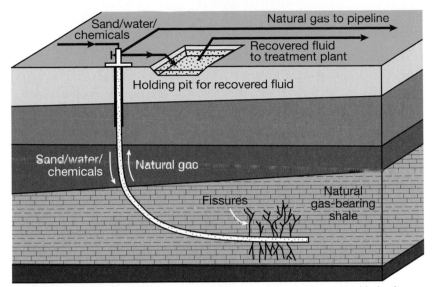

Fig. 5–20. Hydraulic fracturing an oil well (*Source:* Adapted from graphic by A. Granberg for Pro Publica Inc., Nov. 13, 2008)

After the injection phase is completed, the fluid and any loose proppant are back-flushed out of the well so as not to impede production. A portion of the frac fluid is generally recovered and stored in pits or containers; then, it may be processed to allow reuse in later fracing operations. Some of the fluids are treated (cleaned) and then released to the environment, while some other residual material is placed in permanent deep-well storage.

Results. Experts estimate that hydraulic fracturing can increase a well's initial production rate by 1.5–30 times, with the greatest increases seen in low-permeability (tight) reservoirs. In fact, some tight formations may produce for no more than a few months (or not at all) if not stimulated in this way.

A well can be fraced several times during its life, and the technique may increase ultimate production by 5%–15%. This technique has been used with great success since about 2006 to increase gas production from shale formations across the United States, including the Barnett (Texas), Haynesville, and Fayetteville (southeast), and Marcellus (northeast).

In an April 2009 report, DOE reported shale gas production of only about 0.26 tcf in 1998.[5] That figure jumped to 1.4 tcf in 2007, and DOE expected production of 4.8 tcf in 2020. DOE stated that shale gas production potential of 3–4 tcf per year may be sustainable for decades. (Roughly, 1 tcf is enough gas to heat 15 million homes for one year.)

In 2007, engineers found they could use the method to extract oil from the large Bakken formation beneath North Dakota and Montana (fig. 5–21) by increasing the number of cracks in the rock and using different chemicals. In 2010, production there topped 458,000 b/d.

Fracing is now also being applied in the Eagle Ford oil shale formation of South Texas. By some estimates, the Bakken and the Eagle Ford are each expected to ultimately produce four billion barrels. That would make them the fifth- and sixth-biggest oil fields ever discovered in the United States. The technique is also being considered for use in newer fields in the Niobrara (under Wyoming, Colorado, Nebraska, and Kansas), Leonard (under New Mexico and Texas), and Monterey (under California) formations.

EIA estimates oil production from shale will grow to about 500,000 b/d by 2015. Some oil executives and analysts say that figure could even reach 2 million b/d.[6]

Several foreign oil companies (e.g., Shell, BP, and Statoil) are pursuing investment in U.S. oil shale development. In particular, the China National Offshore Oil Corporation, China's national oil company, signed two deals in late 2010 and early 2011 (together worth $1.5 billion) with Chesapeake Energy for stakes in shale projects in the Niobrara and Eagle Ford formations.

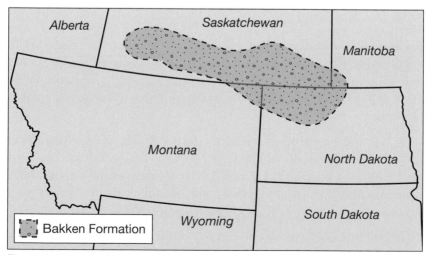

Fig. 5–21. Bakken shale formation map (*Source:* Adapted from Geology.com)

Environmental concerns. Despite this record of extensive use, environmental groups and others have expressed concerns about the environmental impact of hydraulic fracturing and the management of the volumes of water used to fracture shale. As of mid–2011, a number of states (notably in the Northeast, where development of the Marcellus gas shale resource is under way) were considering legislative action to oppose hydraulic fracturing. In addition, several class-action lawsuits were filed against gas developers.

One specific concern has been possible contamination of groundwater by either the materials used for fracturing or by the methane released as a result. Industry experts point out that wells are typically drilled to depths of 6,000–10,000 feet, well below the water table and separated from it by many tons of impermeable rock, but concerns persist.

Regarding fluid composition, the gas industry points to the 2009 DOE study that describes a typical mixture of frac fluid as 95% water and 4.5% sand, with just 0.5% made up of chemical additives.[7]

Well operators consider the composition of many frac fluids to be proprietary information, but they do make details available to regulators, first responders, and medical personnel for use in protecting human health and the environment. For example, since July 14, 2010, well completion reports by Range Resources (based in Fort Worth, Texas) have included detailed information about all additives the firm uses in its hydraulic fracturing operations in the Marcellus Shale. The data go to Pennsylvania state environmental officials and are posted on the company's Web site.[8]

One provision of the federal Energy Policy Act of 2005 exempted hydraulic fracturing operations from compliance with the Safe Drinking Water Act. Nevertheless, well operators must comply with a range of federal chemical record-keeping and reporting requirements and with individual state laws and regulations concerning fracturing operations.

Federal legislation was proposed in 2008 and 2009 as the Fracturing Responsibility and Awareness of Chemicals Act (FRAC Act). Reintroduced in 2011, it would give the federal Environmental Protection Agency (EPA) the authority to regulate hydraulic fracturing in some states. However, as of February 2012, lawmakers had taken no action on the proposed law.

EPA released a study in 2004 that showed hydraulic fracturing did not pose a threat to drinking water supplies. However, in early 2011, the agency submitted a draft plan for further study to the agency's independent Science Advisory Board for its review. EPA expects release of initial study results by the end of 2012. An additional report based on further research is planned for release in 2014.

In April 2012, the EPA updated standards for oil and gas production that will reduce methane and other gaseous emissions released into the air from *all* new gas wells (effective Jan. 2015) that are hydraulically fractured.

Several European countries are also weighing the benefits and impacts of hydraulic fracturing. In May 2011, France imposed a moratorium on the practice, but a committee of the British parliament judged fracing to be acceptable. Poland, which may have Western Europe's greatest shale gas reserves, had awarded dozens of concessions for development as of mid–2011, but was proceeding carefully.

References

1   McFall, K. S., D. E. Wicks, and V. A. Kuuskraa. 1986. A geologic assessment of natural gas from coal seams in the Warrior Basin, Alabama. Topical report 86/0272. Des Plaines, IL: Gas Research Institute.

2   Ely, J. W., R. L. Tiner, M. Rothenburg, A. Krupa, F. McDougal, M. Conway, and S. Reeves. 2000. GRI's restimulation program enhances recoverable reserves (parts 1 and 2). *World Oil*. November.

3   Reeves, S. R., and D. E. Wicks. 1994. Stimulation technology in the Antrim shale. Topical report 94/0101. Des Plaines, IL: Gas Research Institute.

4   Borchardt, J. K. 2010. 1300 Fathoms. *Mechanical Engineering*. December.

5   Ground Water Protection Council and ALL Consulting. 2009. Modern shale gas development in the United States: A primer. Prepared for the U.S. Department of Energy Office of Fossil Energy and the National Energy Technology Laboratory.

6   Fahey, J. 2011. New drilling method opens vast oil fields in US. Associated Press. Published February 9, 2011, in the *Denver Post* (and other channels).

7   Ground Water Protection Council and ALL Consulting. 2009.

8   Range Resources Corp. 2010. Voluntary Well Completion Reports. Posted on company Web site: www.rangeresources.com. Fort Worth, TX. (Accessed Feb. 13, 2012.)

6 Drilling and Completion: Offshore

Chapter 5 described onshore (land) drilling rigs. However, substantial amounts of oil and gas are also found off the coasts of many countries— often in deep water and deep below the seabed.

Although the general approaches for drilling into the seafloor and completing a well are much the same as those used when drilling on land, special vessels, equipment, platforms, and procedures are needed. This chapter will focus on offshore drilling and completion operations.

Exploratory Drilling

Moveable rigs are most often used for exploratory offshore drilling because they are much cheaper to use than permanent platforms. Moveable rigs include drilling barges, drill ships, jack-up rigs, and semisubmersible rigs.

Once large deposits of hydrocarbons have been found, permanent platforms of various designs are typically built for development drilling and production.

One important difference between offshore and onshore drilling rigs is the manner in which rotary power is transmitted to the drill string. Offshore rigs use a device called a *top drive,* instead of the drawworks and associated cables used on a land rig. The top drive is a large electric or hydraulic motor, typically 1,000 horsepower or more, that hangs near the top of the drilling derrick from the hook at the bottom of the traveling block. The top drive turns a shaft into which drill string sections can be screwed, commonly three sections at a time.

Drilling barge

For exploratory drilling in relatively shallow water (as deep as about 25 feet), a drilling barge can be used. One common design is a *posted* barge, designed to float to the site of interest and then be sunk, resting on the bottom for stability. The actual drilling platform is raised on posts extending upward from the deck of the barge, so that the drilling deck is above the water's surface.

Mobile offshore drilling unit

Further offshore, in deeper water, a mobile offshore drilling unit (MODU) is used. The three types of MODU are the jack-up rig, the semisubmersible rig, and the drill ship.

Jack-up rig. The jack-up is the most common, generally used in water as deep as 400 feet (fig. 6–1). It consists of two bargelike hulls (one above the other) plus three or four vertical legs that pass through the hulls. The drilling platform and derrick are mounted on the upper hull. The whole unit is usually towed into position by tugboats, with the two hulls fastened to each other and the legs jutting into the air (as high as 600 feet above the water).

Once in position, the lower hull, or *mat*, is slowly flooded while rack-and-pinion mechanisms simultaneously jack each leg downward. When the mat is securely resting on the seabed, the rack-and-pinion units then jack the platform upward, well above the surface to eliminate the impact of waves and tides. On most jack-up rigs, the drilling derrick is cantilevered—mounted on two large steel beams that extend over the edge of the deck. In other cases, the derrick is mounted closer to the center of the platform and drills down through an opening in the deck.

Once drilling is completed, the two hulls are brought back together and the legs again raised into the air. The entire rig is then towed to another site.

Semisubmersible rig. As the name implies, a semisubmersible rig, or *semisub*, is partially submerged to increase stability. The unit is typically towed to the drilling site. The drilling platform and derrick are connected by columns to pontoons that extend 30–50 feet below the water's surface (fig. 6–2).

Fig. 6–1. Jack-up rig (*Source: Oil & Gas Journal*, March 3, 2008)

Fig. 6–2. Semisubmersible rig (*Source: Oil & Gas Journal*)

Ballast water is pumped into and out of the pontoons to achieve the desired stability. In relatively shallow water, a system of chains and anchors hold the floating rig in place. In deeper water, a dynamic positioning system (using Global Positioning System satellites, computers, and shipboard thrusters) keeps the vessel in the desired spot. Some semisubs can drill in water as deep as 10,000 feet.

Drill ship. The third type of MODU is the drill ship, a large vessel specially modified for drilling (fig. 6–3). The derrick is mounted in the center of the ship, and the operators drill down through a hole in the hull. As with the semisub, the drill ship uses a dynamic positioning system to keep it on station. Some drill ships have two drilling rigs.

Fig. 6–3. Drill ship (*Source: Oil & Gas Journal*)

Transocean reported in April 2011 that its ultra-deepwater drill ship *Dhirubhai Deepwater KG2* claimed a world record for operation in the greatest water depth by an offshore drilling rig—10,194 feet—while working for Reliance Industries offshore India. The vessel is equipped to work in water as deep as 12,000 feet and is outfitted to construct wells as deep as 35,000 feet.

Well Completion

For a well drilled by a jack-up rig (in relatively shallow water), several hundred feet of large-diameter (26- or 30-inch) conductor casing is set into the seafloor. This can be accomplished by using water jets or a pile driver (in relatively soft soil) or by drilling a hole (in hard rock), running the casing into the hole, and then cementing it in place. This conductor casing extends up out of the water to a level just below the drilling deck.

A smaller-diameter hole is then drilled down the centerline of the casing to several hundred feet below the bottom of the conductor casing. The next section of surface casing (slightly smaller in diameter) is then run into that hole and cemented. A BOP system is then bolted to the top of the surface casing, and the rest of the well is sequentially drilled and cased in a manner similar to that used for onshore wells.

Completion of wells drilled by a semisubmersible rig or a drill ship (in deeper water) is somewhat more complex. The drill string itself is used first to lower a temporary guide base—a hexagonal steel framework—and then a guide frame to the seabed. Those two units are used to position the drill bit, and a hole 30 or 36 inches in diameter is drilled to about 100 feet.

The drill string and guide frame are then raised to the surface, and attention turns to the *foundation pile*—the first string of casing that will be placed in the hole. The guide frame is attached to the bottom of the foundation pile, and a permanent guide structure is attached to the top.

The whole assembly is lowered to the seafloor (using steel cables that run from the ship to the temporary guide base), and the foundation pile is run into the drilled hole and cemented. The permanent guide base (now flush with the seafloor) is mated with the temporary guide base.

The drill string is lowered again, the hole is drilled deeper, and a string of conductor casing is then cemented in place. A subsea BOP is lowered and locked onto the wellhead. Then, a flexible, hollow metal tube—a marine riser—is attached to the BOP stack and to the drilling rig. Once the drill string is inserted into the marine riser, drilling mud can be directed down through the drill string, returning upward through the annulus between the drill string and the wall of the marine riser.

A variety of special equipment is used to compensate for the motion of the semisubmersible or drill ship due to wind and waves. In the event of severe weather or other emergency, the BOP can be closed and the stack disconnected from the marine riser, allowing the ship to move to safety. Later, the connection can be reestablished.

The wellhead equipment on a completed subsea well is similar to that on onshore wells. In some cases, the subsea wellhead is *dry* (enclosed in chamber maintained at atmospheric pressure). In other cases, the equipment is *wet* (exposed to seawater).

Production from a subsea well can be carried by flowline to a manifold where it is combined with production from other wells (fig. 6–4). From the manifold, oil typically flows to a production platform (see next section) or to a floating production, storage, and off-loading vessel (FPSO). The FPSO is a converted or custom-designed ship with onboard facilities for oil separation and treating. Oil from the FPSO is often then transferred to a shuttle tanker that takes it to an onshore facility for further treatment or storage.

In some cases, undersea pipelines are constructed to carry production from a platform to shore.

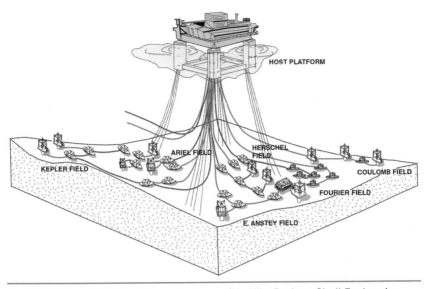

Fig. 6–4. Subsea field development scheme (Na Kika Project, Shell Exploration and Production) (*Source:* Leffler, Pattarozzi, and Sterling)

Fixed Development and Production Platforms

If exploratory drilling successfully identifies promising hydrocarbon deposits, then an operator decides either to plug the well and defer production to a later date or to proceed with development and production. Large, complex offshore platforms are used when the decision is made to proceed. Both fixed and floating platform designs are used, depending primarily on water depth.

There are four main types of fixed platform used for offshore oil and gas production. These are the steel–jacket, the gravity-based, the tension-leg, and the compliant-tower designs.

Steel-jacket platform

The steel-jacket platform has four to eight legs of welded steel pipe, which comprise the *steel jacket*, stabilized with multiple crisscross truss members (fig. 6–5). Typically weighing about 20,000 pounds, the platform is built on land, carried by barge or floated (in a horizontal position) to the site, and then rotated vertically and sunk into position. The bottom of each leg is welded, bolted, or cemented to piles driven into the seafloor.

Once the platform is securely in place, an onboard crane lifts the drilling deck, crew quarters, and various modules (e.g., for power generation or mud storage) onto the platform from barges. The steel jacket platform is generally used in water about 500 feet deep.

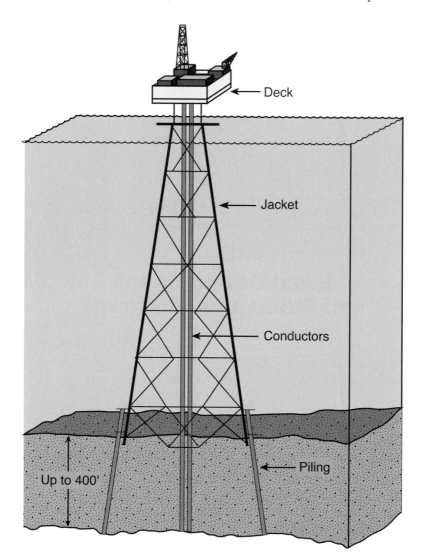

Fig. 6–5. Steel-jacket offshore platform (*Source:* Adapted from Leffler, Pattarozzi, and Sterling)

Gravity-based platform

The gravity-based platform design is also typically used in water as deep as about 500 feet deep. One or more reinforced concrete columns, or *caissons*, support the work platform that is mounted on top of them (fig. 6–6). Hollow chambers in the columns, plus ballast tanks mounted at the bottom of the columns, can be filled with fluids or solid material to achieve the desired degree of buoyancy.

Fig. 6–6. Gravity-based offshore platform (*Source:* Leffler, Pattarozzi, and Sterling; courtesy Norske Shell)

After the structure is built at a sheltered port, the buoyancy of the base is adjusted so that the entire structure floats and can be towed to the work site. There the base is filled with ballast, lowering it to the seafloor and leaving only the topsides visible above the water. During operation, crude oil can be stored in the hollow chambers of the submerged columns.

As the name implies, gravity holds the massive base in place on the seafloor, with no need for pilings or anchors. This design is effective in areas with rough seas. For example, the *Hibernia* platform of Mobil Oil— about 200 miles off the coast of Newfoundland, in 260 feet of water— includes a concrete gravity-based structure weighing about 450,000 tons.

Tension-leg platform

In the tension-leg design (fig. 6–7), long, hollow steel tubes, typically two feet in diameter, are connected to the bottom of the platform deck. These tubes, called *tendons*, extend down to the seabed, where large weights hold them in place, or they are bolted to pilings sunk into the seabed. The tendons and weights prevent the platform from moving up and down because of wave or tidal forces.

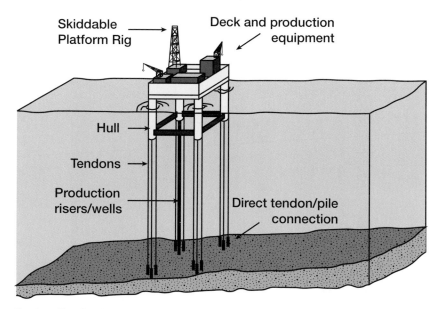

Fig. 6–7. Tension-leg offshore platform (*Source:* Adapted from Leffler, Pattarozzi, and Sterling; courtesy Shell Exploration and Production)

Compliant-tower platform

A final type of fixed platform is the compliant tower (fig. 6–8), with a subsea design structure that is less massive and is more flexible than those of other rigid fixed-platform designs. With its smaller "footprint," the compliant tower is suitable for installation in water 1,000–3,000 feet deep without any need for guy wires or heavy anchors.

The deck on which the drilling platform rests is attached to a strut-and-truss structure (similar to that of a steel jacket platform) that is in turn connected to a set of special-design pilings driven into the seafloor.

The pilings are specifically designed to be somewhat flexible rather than rigid. Working together, they allow limited platform sway in reaction to wind and waves but then restore proper alignment quickly.

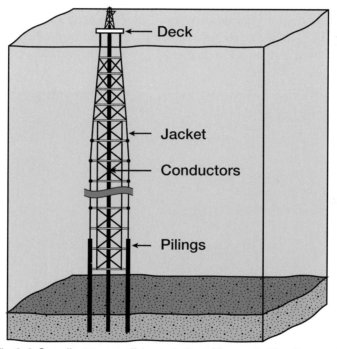

Fig. 6–8. Compliant-tower offshore platform (*Source:* Leffler, Pattarozzi, and Sterling)

Floating Development and Production Platforms

There are three main types of floating platform: the semisubmersible vessel, the monohull vessel, and the spar platform. (Note that tension–leg platforms are also used for some deepwater projects.)

Semisubmersible vessel

As described earlier, semisubmersible vessels can be used for exploratory drilling. However, much larger and more–complex semisub units also have been built for use as platforms for development drilling and production operations in deep water. These vessels offer greater stability than so-called monohull vessels of conventional shiplike design.

On a deepwater semisub, structural columns that support the work platform well above the water's surface are also connected to a submerged hull (fig. 6–9). This hull consists of pontoons or similar structures to which ballast can be added or removed. Stability is maximized when the hull is fully ballasted and at its lowest depth.

Fig. 6–9. Semisubmersible offshore platform (*Source:* Leffler, Pattarozzi, and Sterling)

Semisubs are commonly towed into position by tugboats. Once in place, the vessels can be kept in precise position with dynamic positioning systems (described earlier).

Thunder Horse PDQ—the largest semisub in the world, displacing 129,000 tons—is moored in 6,050 feet of water in the Gulf of Mexico, 150 miles southeast of New Orleans. Operated jointly by BP and ExxonMobil, it began oil production in 2008. Four columns 118 feet tall connect the production platform to the submerged hull.

Monohull platform (FPSO)

An FPSO is a widely used type of monohull platform (fig. 6–10). From a distance, it can be hard to distinguish from an oil tanker. It is designed to receive hydrocarbons produced from one or more subsea wells or from nearby platforms. FPSOs are attractive for use in frontier offshore regions because they can be brought to a site easily and do not require the presence of a local oil pipeline infrastructure.

Aboard the FPSO, the oil undergoes initial processing to remove water and other unwanted components and is then pumped into storage compartments in the hull. (Some FPSOs are also equipped to receive, process, and store natural gas.) Processed oil is periodically transferred (off-loaded) to a shuttle tanker or a barge or is pumped to shore through a subsea pipeline.

Fig. 6–10. FPSO (monohull platform) (*Source: Oil & Gas Journal*)

An FPSO can be a converted oil tanker or a custom-built vessel. Alternatively, depending on expected operating conditions, an FPSO may look more like a simple box than a ship, may be a semisubmersible with storage, or may even have a cylindrical shape. If oil is not processed onboard, then the vessel is considered an *FSO*—floating storage and off-loading vessel.

In most cases, the lines (risers) carrying oil from wells or nearby platforms are connected to a turret at the front of the FPSO. The turret lets the vessel rotate to head into the wind and reduce mechanical forces on the moorings. Some turrets and mooring systems are designed to be disconnected if necessary; others are permanently connected. In the often stormy North Sea, most ship-shaped FPSOs are purpose-built and permanently moored.

The world's first *FDPSO*—floating drilling, production, storage, and off-loading vessel—was developed in 2009 for Murphy Oil, for operation off the coast of the Republic of Congo. The *Azurite*, owned by Prosafe Production, incorporates deepwater drilling equipment that can be removed for later use at other sites. The unit is a converted very large crude carrier (VLCC).

Spar platform

The last major type of offshore production platform is called a *spar*. In this design, the work platform is built atop a tall cylinder that is designed to be submerged in the vertical position below the platform. Spars are built in three configurations (fig. 6–11):

- conventional spar, with a one-piece cylindrical hull
- truss spar, with a midsection composed of truss elements that connect an upper buoyant hull with a lower tank containing permanent ballast
- cell spar, built up from several vertical cylinders

Inherently more stable than a tension leg platform because of the heavy ballast in its bottom section, the spar also is attached to the seafloor with mooring lines. The lines can be adjusted to move the spar horizontally (within limits) to better position it above wells on the seabed. *Strakes* (resembling fins) installed on the outer surface of a cylindrical spar can help lessen the effects of wave action and water movement on the platform.

In March 2010, oil production began from the world's deepest production platform, the *Perdido* spar, operated by Royal Dutch Shell.

The $3 billion facility is sited 200 miles from Galveston, Texas, in 8,000 feet of water in the Gulf of Mexico. The spar measures 188 feet in diameter and 555 feet in height, incorporating both a cylindrical section (topmost) and a lower truss structure. Nine mooring lines, each two miles long, are attached to the base of the truss. Oil and gas will be produced from 35 wells below the platform, some extending 14,000 feet below the seabed.

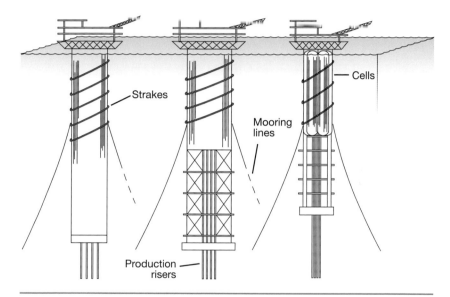

Fig. 6–11. Spar platform options (*Source:* Leffler, Pattarozzi, and Sterling)

7 Managing Oil and Gas Production

Once drilling has been completed, a major business objective for the exploration and development firm is to effectively manage production to achieve cost-effective, safe, and environmentally sensitive extraction of crude oil and natural gas from subsurface reservoirs. This chapter focuses on actions that optimize production from oil and gas wells, including maintenance, repair, and improved-recovery programs that help achieve this goal while extending the productive life of wells. The initial sections present information pertinent to both oil and gas wells; those sections that come after focus first on oil production and then on gas production.

Start of Production

As production operations begin, the rig and other equipment used to drill and complete the well have been moved away from the wellbore, and (as noted in chap. 5) the top is usually outfitted with a collection of valves and other components called a *Christmas tree*. These valves regulate pressures, control flows, and allow access to the wellbore in case further completion work is needed. From the outlet valve of the production tree, the flow can be connected to a distribution network of pipelines and tanks that eventually moves the product to refineries, natural gas compressor stations, or oil export terminals.

As long as the pressure in the reservoir remains high enough, the Christmas tree is all that is required to produce the well. If well pressure drops, artificial lift methods can be employed to maintain economic production.

Artificial Lift

The natural pressure within any liquid-producing reservoir will force fluid (liquid or gas) to areas of lower energy or potential. When the pressure inside a production well is below the reservoir pressure, this differential will push fluids into (and upward within) the wellbore. This is typically the case for gas wells; it can also be true for some oil wells during early production.

For many oil wells, however, reservoir pressure may not be sufficient to push the fluid to the surface, creating the need for some type of artificial lift. The lift method chosen depends on the depth of the reservoir (greater depth results in a higher pressure requirement) and the density of the fluid (a heavier mixture results in a higher requirement). There at least a dozen artificial lift approaches in use by industry.

Following initial production, two things happen: water (which is heavier than oil and much heavier than gas) begins to encroach into the formation; and reservoir pressure drops as the reservoir depletes. If no action is taken, then flow from the well will eventually stop. However, most well operators will implement an artificial lift program to continue or increase production.

The most common approach is to remove the Christmas tree and install a surface pump. The motor-driven sucker-rod pump (also called a *pumpjack*) is by far the most common. From a distance, this arrangement resembles a large bird or a horse's head slowly bobbing up and down (fig. 7–1).

Another approach, called *gas lift*, injects a compressed gas (often, natural gas) into the annulus between the casing and the tubing. Special valves allow the gas to enter the tubing, where it dissolves in the produced gas/water liquid and also forms bubbles. The dissolved gas and bubbles force the liquid up the tubing string, where the gas can be captured and recycled.

A third alternative is to install a powered pump—electric or hydraulic—at the bottom of the well tubing. In one case, power is delivered by an electrical cable running down the well; in the other, high-pressure hydraulic fluid is delivered to the pump to drive it.

A fourth method is to pump hydraulic fluid down into the tubing string itself, allowing the produced gas and water to flow upward through the tubing/casing annulus to the surface. Other approaches are aimed at removing water from gas wells, including:

- Dropping soap sticks (typically of 2-inch diameter, 16 inches long) downhole to mix with water and create a lighter-weight foam that can be pushed upward more easily and steadily by reservoir pressure than intermittent heavy slugs of water

- Installing smaller-diameter tubing sections (called *velocity strings*) whose smaller cross-sectional area increases gas velocity, reestablishing the critical speed needed for the gas to carry water along with it up the wellbore

Fig. 7–1. Sucker-rod pumping unit (*Source:* Hyne, N. J.)

Workover of a Well

After months or years of oil and gas extraction, remedial action may be needed to maintain production from a well. Several actions collectively called a *workover* are typically performed by contractors by using special skills and equipment.

Prior to workover, a well must be shut in to stop production. Flow is killed by pumping brine, drilling mud, oil, or some other fluid down into the annulus, between the casing and the tubing, and back up the tubing string. A BOP is usually installed as well, and the tubing string and other downhole equipment are removed.

Workover actions

Workover can include the following steps:

- Removal of water or drilling mud (swabbing), done either right after a well is completed or later (to restore production in an operational well)

- Repair or replacement of downhole pumps, valves, and packers

- Installation of smaller-diameter tubing to boost flow rates

- Removal of scale (salts, e.g., calcium sulfate) or paraffin from tubing

- Cleanout of loose sand from the bottom of wells drilled in sandstones

- Repair of damaged tubing or casing or of the cement around the casing

Sidetrack well

In some situations, the well casing may be significantly damaged or may have otherwise lost integrity. Replacing the cemented casing is expensive, so one alternative is to drill and complete a sidetrack well. A hole is made in the casing above the obstruction or damage, and the well is plugged with cement below that hole. Special drill tools—such as a whipstock, bent housing, or bent sub (see chap. 5)—are used to drill outward at an angle from the original well. The new wellbore is completed in the conventional manner, and drilling continues after a liner is set.

Recompletion

If operators conclude that a particular pay zone is depleted, a technique called recompletion may be undertaken. This is the completion of other pay zones below or above the original zone. Cement is used to seal off the original producing zone.

Workover of offshore wells

In offshore wells, workover can be very expensive because of the need to shut in production, as well as to hire special-purpose barges or platforms and trained staff to conduct the work. Depending on water depth, some work can be done by divers; in deep water, however, use of a remotely operated vehicle (ROV) is often a better option (fig. 7–2). The ROV is tethered to the production platform, and its motion and depth, as well as its lights, robotic arms, and specialized tools, are controlled by an operator at the surface.

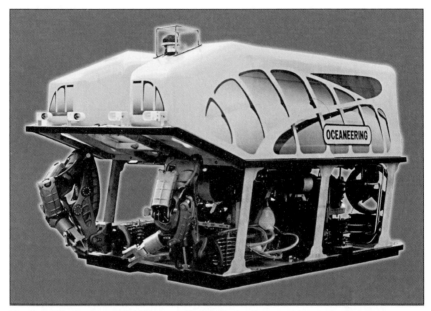

Fig. 7–2. Work-class ROV (*Source:* Hyne, N. J.; photograph courtesy Oceaneering International)

Improved Oil Recovery

Oil produced by the original pressure in a formation is called primary production. For most wells, primary production yields only about 30%–35% of the oil in place. (For gas wells, primary production generally yields about 80% of the gas in place, so less attention is given to improved recovery.) To extract additional oil, operators have developed a range of methods to inject various materials into a reservoir, aiming to push remaining oil toward wellbores.

Waterflood

A common technique often used first by operators to extract additional oil is called *waterflood* (fig. 7–3), which can recover 5%–50% of the remaining oil. They introduce water (treated first to remove solids, bacteria, organic matter, and oxygen) through several injection wells placed in specific patterns around a producing well.

A waterflood program must be matched to the composition, permeability, and degree of homogeneity of the target rock to avoid damaging the formation and to maximize oil recovery. In some cases, polymers (to increase viscosity) or chemicals (to modify acidity) are added to the water.

Another set of methods to extract additional oil are collectively known as *enhanced oil recovery*. These methods involve the introduction of substances not naturally found in a reservoir and include the use of gases, chemicals, or thermal energy. The aim is to boost reservoir pressure and to sweep hydrocarbons from various parts of the reservoir toward wellbores.

As with waterflood, these methods require injection wells (often chosen from old production wells in a carefully determined pattern). They are also used when facing such problems as reservoir pressure depletion or high oil viscosity.

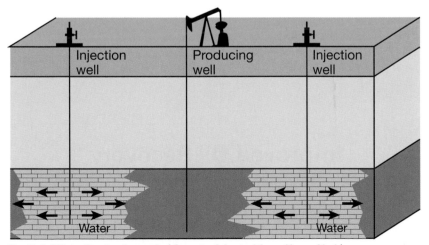

Fig. 7–3. Waterflood schematic (*Source:* Adapted from Hyne, N. J.)

Gas injection

This technique typically sends carbon dioxide, nitrogen, or liquefied petroleum gas (LPG) into the reservoir—gases that are miscible (dissolve in the oil). The gas then pushes the fluid oil through the pores and cracks in the rock toward producing wells. This approach can often recover 35% of remaining oil.

Chemical flood

This approach injects different chemicals—each serving a different purpose—into a depleted sandstone reservoir in separate batches, or *slugs*. A common first step is to inject a slug of a detergent-like surfactant that reduces the surface tension of the oil, which washes out of the rock pores and forms a micro-emulsion. The next slug—water thickened with polymers—pushes the micro-emulsion toward the producing wells. A chemical flood can recover about 40% of remaining oil.

Thermal recovery

Several types of thermal recovery are used to make heavy oil remaining in a reservoir flow more easily.

Cyclic steam injection. One common approach is to use cyclic steam injection (fig. 7–4). Operators inject steam into a well and then shut in the well for as long as two weeks, to allow the heat from the steam to boost the temperature of the oil and make it flow more easily. After that, a surface pump pulls the heated heavy oil from the well over a period of days or weeks. This "puff and huff" cycle of injection and pumping is typically repeated until it becomes ineffective (perhaps 15 or 20 times).

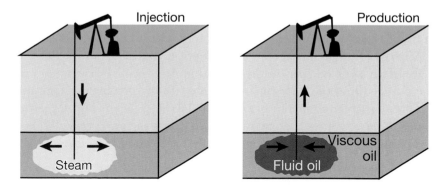

Fig. 7–4. Cyclic steam injection (*a*) and production (*b*) (*Source:* Adapted from Hyne, N. J.)

Steamflood or steam drive. Another thermal recovery method is the steamflood or steam-drive technique, which is similar to the waterflood method described above. Superheated steam is pumped into injection wells to heat the heavy oil, reducing its viscosity. In addition, as the steam condenses, the resulting hot water pushes the oil toward separate producing wells.

Wells used for a steamflood are similar to those used in the waterflood method described above, but they are spaced more closely. This method can recover 25%–65% of the oil in place.

Steam-assisted gravity drainage. A variation of steamflooding, *steam-assisted gravity drainage* (SAGD) has been used with great success in the tar-sands formations of Alberta. Steam is injected into an upper horizontal well to melt the tarlike bitumen, which flows downward into a lower horizontal well. The bitumen is then pumped from that lower well to the surface. By some estimates, SAGD can recover up to 60% of the oil in place.

Alberta oil-sands projects

In a December 2010 inventory of major oil-sands projects, the Alberta government listed 55 in various proposal or construction stages. The report also indicated that bitumen production totaled about 1.5 million b/d in 2010 and that companies had produced more than 7 billion barrels as of January 2011.

Fireflood or in situ combustion. A third kind of thermal recovery is a fireflood, or *in situ combustion* (fig. 7–5). In relatively shallow reservoirs, operators use a gas burner or other ignition source to set on fire the subsurface oil in one part of the formation. In deeper reservoirs, air can be pumped into the reservoir to start the fire by spontaneous ignition. Thereafter, large volumes of air are injected to keep the fire burning. Heat from the fire reduces the viscosity of the oil; at the same time, the hot gases generated by the fire push the heated oil toward producing wells.

In some cases, water is injected along with the air, or water and air are injected alternately. In either case, the water turns to steam within the reservoir, creating additional driving pressure to move the oil. The fireflood method can recover 30%–40% of the oil in place.

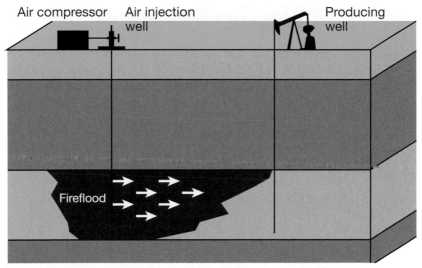

Fig. 7–5. Fireflood schematic (*Source:* Adapted from Hyne, N. J.)

Gas Production Management— an Overview

The processes used to produce gas from an underground reservoir depend in large measure on the type of well. Natural gas from a well that was drilled chiefly to extract oil is called *associated gas*—that is, the gas is associated with oil production. In some cases, associated natural gas is pumped back down into the formation to provide pressure to aid in further oil extraction. However, it may also exist in sufficient quantities to be worth producing and processing along with the oil.

If a well is drilled specifically for natural gas and yields little or no oil, then the gas produced is called *nonassociated gas*. As might be expected, it is generally cleaner than associated gas and requires less subsequent processing.

Condensate wells contain natural gas, as well as a liquid hydrocarbon mixture (condensate) that is often separated from the natural gas either at the wellhead or during later processing of the gas.

If tests indicate that a well is likely to be a satisfactory producer, then the well is completed (as described in chap. 5). Not only does production begin, but a series of treatment steps are taken to improve gas quality, to reach a level acceptable to shippers and buyers.

Initial Gas Treatment

About 75% of the raw natural gas in the United States comes from underground reservoirs that hold little or no oil. This nonassociated gas is cleaner (and therefore easier to process) than gas from wells that do contain oil. But regardless of the source, dirt, sand, and water vapor must be removed from the raw gas to prevent contamination and corrosion of equipment and pipelines.

A device called a wellhead separator performs initial cleanup, removing water, condensate and sediment. Dirt and sand are removed with filters or traps. Water vapor is typically removed by passing the gas through a desiccant material, such as silica gel or alumina (solid granules) or glycol (a liquid).

Nevertheless, the gas output from the wellhead separator can still contain hydrocarbons, hydrogen sulfide, and a range of noncombustible gases. Further treatment is needed, and this is most cost-effectively done at a large central facility called a natural gas processing plant or simply a gas plant (fig. 7–6). A network of small-diameter, low-pressure piping called a gathering system routes the gas from multiple (sometimes more than 100) wells to the gas plant.

Fig. 7–6. Williams Companies Rio Blanco gas plant (Colorado) (*Source: Oil &Gas Journal*)

Gas Processing Operations

At the gas plant, initial steps remove any remaining particulate matter and further reduce the water–vapor content of the incoming gas. Then, focus turns to the separation of hydrocarbons and fluids to yield pipeline-quality dry natural gas (fig. 7–7).

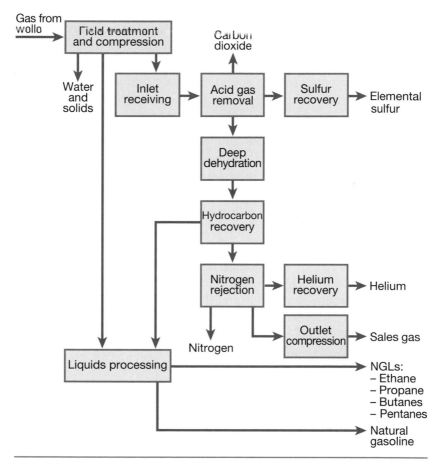

Fig. 7–7. Gas plant schematic (*Source:* Adapted from Kidnay, Parrish, and McCartney)

Removal of carbon dioxide and hydrogen sulfide

Natural gas that contains significant amounts of carbon dioxide or hydrogen sulfide is sometimes called acid gas because these two compounds can react with water vapor in the natural gas to form acids that cause corrosion. (When only hydrogen sulfide is present, natural gas is referred to as *sour*. Natural gas is described as sweet after the removal of hydrogen sulfide.)

A wide range of technologies have been developed to remove acid gases, including use of chemical, physical, and hybrid absorption solvents; molecular sieves and other solid adsorbents; membranes; direct chemical conversion processes; and cryogenic (ultralow-temperature) fractionation. One common approach flows raw natural gas up through a tower while a spray of water mixed with a solvent is injected at the top. The solvent reacts with the two gases (carbon dioxide and hydrogen sulfide), and the solution is drained from the bottom of the tower for further processing.

When present in sufficient amounts, carbon dioxide can be injected into an underground formation under pressure to improve oil recovery. Sulfur can be converted to elemental sulfur for commercial sale or can be injected into a suitable underground formation.

Removal of water vapor

Dehydrators reduce the moisture content of natural gas by the use of either solid drying agents or, more commonly, water-absorbing liquids. Besides affecting the heating value of the natural gas and contributing to corrosion, water can (under certain pressure and temperature conditions) become part of a solid or slushlike compound, methane hydrate. A lattice structure of methane and water molecules, hydrate can plug pipelines, pressure regulators, and other equipment.

Removal of heavier hydrocarbon gases

Raw (but cleaned) natural gas can still contain significant amounts of heavier hydrocarbon gases (such as propane and butane) that can be extracted and sold separately. The most common extraction method is to bubble the raw gas up through a tower containing a cold absorption oil, similar to kerosene.

As the gas comes in contact with the cold oil, the heavier hydrocarbon gases condense into liquids and are trapped in the oil. The lighter hydrocarbon gases, such as methane and ethane, do not condense into liquid and flow out the top of the tower. About 85% of the propane, 40% of the ethane, and almost all of the butane and heavier hydrocarbons can be extracted using this absorption process. The absorption oil is then itself distilled to remove the trapped hydrocarbons, which are separated into individual components in a fractionation tower (see chap. 11).

After this treatment, the natural gas (now chiefly methane) may still contain some ethane, as well as small amounts of other constituents, such as nitrogen and helium. A portion of the ethane is sometimes extracted for use as a raw material in various chemical processes. This is done by first reducing the level of water vapor and then subjecting the gas to repeated compression and expansion to cool the ethane and capture it as a liquid.

Nitrogen removal

Nitrogen can reduce the heating value of natural gas (because nitrogen does not burn). After the removal of carbon dioxide and hydrogen sulfide, the processed natural gas goes through a low-temperature distillation process to liquefy and separate the nitrogen. The collective removal of carbon dioxide, hydrogen sulfide, and nitrogen is called *upgrading,* because the natural gas is made cleaner and will consequently burn hotter.

Helium removal

Natural gas is the main source of helium for industrial use in the United States. Extraction of helium gas is done after the nitrogen has been removed, using a complex distillation and purification process to isolate the helium from other gases. Natural gas leaving the gas processing plant is ready for direct use by energy customers or for injection into a transmission pipeline system.

Other Environmental Issues in Gas Production

Two main environmental issues arise in the production of natural gas. The first relates to water quality and the other to air quality.

Both gas and oil extraction bring significant amounts of water to the surface and this *produced water* must be treated and/or disposed of in ways that minimize environmental impact. For example (as noted in chap. 5), hydraulic fracturing operations typically pump millions of gallons of water into the ground, a significant portion of which is then brought back up to the surface. Further, prior to extraction of methane from coal seams, operators must remove water that collects naturally in the pores of the coal. Other gas-bearing formations also may contain considerable volumes of water.

Produced water typically contains dissolved salts, as well as varying levels of hydrocarbons (oil and grease), chemical additives and proppants, sediment, and even something called *NORM*—naturally occurring radioactive materials (e.g., radium-226 or -228). The petroleum industry has developed a range of technologies and procedures to ensure that produced waters can be treated and disposed of properly. For example, the Gas Research Institute (now Gas Technology Institute) developed a *Produced Water Management Handbook* in 2003. Based on interviews with some 250 oil and gas producers, the handbook discusses such practices as well injection, evaporation, surface disposal techniques, recycling and reuse strategies, reverse osmosis, and downhole gas/water separation.

With regard to air quality, methane is a potent greenhouse gas, trapping heat in the atmosphere approximately 20 times more effectively than carbon dioxide. This means that the gas industry pays particular attention to controlling leaks and emissions of methane from production, processing, and transportation operations.

In 1993, the natural gas industry joined with EPA to launch the Natural Gas STAR Program to reduce methane emissions. Program data from EPA indicate that emissions were reduced by more than 820 bcf from 1993 to 2008 (by 114 bcf in 2008 alone).

When Production Ends

After a well has been depleted to the point at which further economic recovery of oil or gas is not possible, it must be plugged and abandoned. This action is required by law, to prevent briny water from the well from polluting groundwater.

The first step is typically to cut and extract the well casing for salvage purposes. Then, a series of mechanical and cement plugs are installed. These seal off all high-pressure and permeable producing zones in the well. A separate cement plug is often installed in the uppermost section of the well to protect any freshwater reservoirs near the surface.

Offshore wells are plugged and abandoned in the same manner as onshore wells, and all subsea equipment is retrieved. When an offshore platform is abandoned, usually only the deck equipment is salvaged. The legs that held up the deck can be cut off at seabed level, and the entire rig can then be towed to shore for salvage or disposal. Alternatively, the structure can be tipped over to lie horizontally on the seabed, often creating an excellent artificial reef for marine life.

8 Transporting Oil

Once crude oil has been produced from an onshore field, it generally undergoes initial processing close to the well site (as described in more detail below) and is then put into tanks for temporary storage (fig. 8–1). After a period of time in storage, crude oil enters the transportation system (fig. 8–2). It may be moved to a refinery (for conversion into a range of products), to a port (for export), or to the site of a customer (e.g., a power plant) who will use the crude directly.

This chapter focuses on the movement of crude by pipeline, although attention is also given to other transport modes, including tanker vessels, railcars, and tank trucks. Product pipelines—those that carry various finished products (gasoline, diesel, and heating oil) from a refinery to distributors or customers—are discussed later, in chapter 11. Offshore (subsea) pipelines are discussed toward the end of this chapter, after the sections dealing with onshore pipelines.

Fig. 8–1. Crude oil stock tanks at a well site (*Source:* Hyne, N. J.)

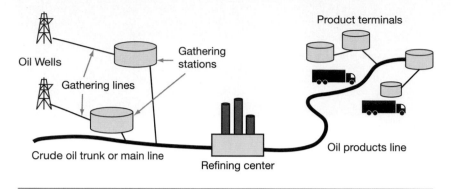

Fig. 8–2. Oil transportation system (*Source:* Adapted from Miesner and Leffler)

Preliminary Crude Processing

A typical land-based oil field contains a number of wells. A network of gathering lines carries this oil from the wells to a location in or near the oil field for preliminary processing.

Oil produced from almost any reservoir contains some gases (including natural gas, often in the form of associated gas) in addition to water vapor or liquid water. For this reason, a unit called a *separator* at a well site is used to remove both the natural gas and the water. Any solid materials (e.g., sand), as well as some dissolved salts, are removed from the raw crude.

Following this initial cleanup, the crude oil plus any condensate (light hydrocarbon liquids extracted during onsite natural gas separation) are pumped into holding (or lease) tanks near the well for temporary storage. (For further information about preliminary cleanup of *natural gas* at or near the wellhead, see chap. 7.)

When a pipeline company is ready to accept the stored oil, a technician or an automated system checks oil volume to be transferred as well as oil quality (to ensure that water and sediment content are below desired limits). In manual operations, after climbing to the top of the storage tank and opening a small hatch, a technician takes three actions:

- Measures the temperature of the oil
- Lowers a gauge tape into the tank to measure oil depth— similar to using a dipstick to check a car engine's oil level
- Takes a sample of the stored oil by dipping a small bottle into the tank or by opening a small valve on the tank's side

Then, by calculation and analysis, the technician will:

- Convert the measured oil depth to oil volume, using a gauge table or computer algorithm
- Separate basic sediment and water (BS&W) from the oil sample, using a centrifuge
- Determine how much BS&W is contained in the measured oil volume.

After performing these actions and calculations (and assuming the BS&W percentage is acceptable to the oil buyer), the technician allows an agreed-on amount of oil to flow out of the tank into a pipeline or tank truck. After withdrawal, the technician measures the oil remaining in the tank, subtracts that figure from the initial measurement, and adjusts the calculated difference in volume for temperature and BS&W content. (Industry standards call for correcting the volume for a temperature of 60°F.)

At some facilities, the oil is measured and tested by skid-mounted *automatic custody transfer* (ACT) units. A technician periodically checks the ACT unit and confirms the BS&W measurement.

After the manual or automated analysis is completed, paperwork is prepared to document the sale. When the oil is delivered to a refinery, a tank farm, or another pipeline, similar analysis and calculation will be conducted to confirm oil quality and delivery volume.

Pipeline Networks

The next sections describe an oil pipeline network, starting with an overview and progressing to a more detailed discussion of the components of a pipeline. Tables 8–1 and 8–2 list major oil pipelines in the United States and in other countries, respectively.

Overview

Long-haul pipelines (also called *mainlines* or *trunk lines*) are larger-diameter pipes with fewer delivery points than short-haul pipelines. The customer generally receives the same quality of oil that was put into the line for delivery but not necessarily the same oil molecules.

The speed of oil flowing through a pipeline depends on pipe diameter, interior-wall condition, pumping-system pressure, topography, and pipeline orientation. In general terms, a given batch of oil moves at 3–8 miles per hour through a mainline, propelled by centrifugal pumps sited every 20–100 miles (depending on terrain).

Table 8–1. Selected major U.S. oil pipelines

Region/name	Route	Comments
Seaway Pipeline	U.S. Gulf Coast to Cushing, OK	
Capline Pipeline	St. James, LA, to Patoka hub in southern IL	
TAPS (Trans-Alaska Pipeline System)	Prudhoe Bay (North Slope) to southern coast of Alaska	Oil then moves by tanker ships to California refineries
Enbridge Pipeline System	Alberta to Chicago and Great Lakes region	Much of the oil is from Alberta tar-sands deposits; 1.5 million b/d
Spearhead Pipeline	U.S. portion of Enbridge system from Chicago to Cushing, OK	
Keystone Pipeline	Hardisty, Alberta, west to Manitoba, south to Nebraska, then east to Patoka hub; also, extension from Nebraska to Cushing, OK	Expansion proposed (Keystone XL) to carry oil from Nebraska to Cushing, OK, and on to Port Arthur, TX.

Sources: Various.

Table 8–2. Selected major oil pipelines outside the United States

Region/name	Route	Comments
Sumed (Suez-Mediterranean) Pipeline	Ain Sikhna terminal on Gulf of Suez (Egypt) to Sidi Kerir on the Mediterranean coast	2.5 million b/d; moves Saudi and other Mideast oil to Europe across Egypt
BTC Pipeline	Baku (Azerbaijan), to Tbilisi (Georgia), to Ceyhan (Turkey)	1.0 million b/d of Caspian Sea oil; by ship from Ceyhan to European markets
Druzhba (Friendship) Pipeline	West Siberia, Urals, and Caspian Sea to Europe	1.4 million b/d
East-West Pipeline	Oil fields in eastern Saudi Arabia to refineries and Yanbu terminal in western Saudi Arabia	
ESPO (Eastern Siberia–Pacific Ocean) Pipeline	Taishet (Irkutsk Oblast, Russia) to Kozimo (near Nakhodka, Russia); also, spur from Skovorodino to Daqing, China, completed Dec. 2010.	1,713 miles, Taishet-Skovorodino (Amur Oblast), done; stage 2 (1,300 miles, to Kozimo) to be done by 2014; 1.6 million b/d to Asian Pacific markets

Sources: Various.

Short-haul lines are also called *spur*, *stub*, or *delivery* lines. These lines usually operate in batch mode, in which the customer receives the same oil molecules that were put into the line. More details about this aspect of pipeline operation are presented in a later section.

Points at which pipelines come together are called hubs (or *marine terminals* if near a port). Significant storage volume is generally available at such facilities.

A major hub for U.S. crude oil pipelines is in Cushing, Oklahoma, approximately 500 miles north of the Gulf of Mexico. This is the hub at which any futures contract for West Texas Intermediate (WTI) crude oil executed on the New York Mercantile Exchange (NYMEX) is physically deliverable. Other primary U.S. hubs and marine terminals are at New York Harbor; the Gulf Coast (Texas–Louisiana coast); Tulsa, Oklahoma; Chicago; and Los Angeles.

At some production sites, no pipeline entry point is nearby. In such cases, trucks, barges, or railcars carry the pipe to another tank facility in an area served by a pipeline. However, pipeline transport is the most common way of delivering crude oil to refineries to be converted into a range of products. Alternatively, the crude oil could be loaded onto a tanker ship for transport to a domestic or foreign refinery or to some other customer. Each of these transport methods is discussed in more detail in later sections.

Pipe

Most sections of an oil pipeline operate at relatively low pressure (up to about 150 psi). Pipe in these sections is made from various types of plastic, fiberglass, or steel. For high-pressure lines (operating pressure sometimes higher than 1,000 psi), steel is used. Some subsea lines are made from more-flexible polymer-and-steel composites.

Pipe for use in cross-country pipelines is manufactured to meet API standards for given applications. These standards specify the physical and chemical properties of both the starting-point metal and the finished pipe. Steel pipe is fabricated with various wall thicknesses and with outside diameters ranging from 4.5 to 48 inches.

A detailed description of the complex process of planning and building an oil or gas pipeline is beyond the scope of the present text. That process includes resolution of numerous financial, legal, environmental, and regulatory issues, as well as the technical challenges of excavation and pipe installation. For the purposes of this book, suffice it to say that long sections (sometimes called *joints*) of steel pipe are moved to the work site and placed into position beside or in a long excavated trench (fig. 8–3). The sections are then welded together either by skilled workers using special equipment at the installation site (fig. 8–4) or by automated welding machines.

Fig. 8–3. Pipeline sections positioned along the route (*Source:* Miesner and Leffler)

Fig. 8–4. Manual pipeline welding (*Source: Oil & Gas Journal*)

Canadian-U.S. Keystone XL pipeline

On January 18, 2012, President Barack Obama made an important decision concerning proposed construction of a new international crude oil pipeline from Alberta, Canada, to the eastern Gulf Coast of Texas. Concurring with a recommendation made in December 2011 by the State Department, he denied a construction permit for the Keystone XL pipeline. (The State Department has authority over the project because the pipeline would cross an international border.)

The Administration concluded that the $7 billion project could not receive an adequate review of environmental, health, and safety impacts by the 60-day deadline set by Congress. However, the action of the president did not preclude later possible approval of the project.

The pipeline, proposed in 2008 by TransCanada, would span approximately 1,700 miles and carry up to 700,000 b/d of oil extracted from Alberta's tar-sands deposits to refineries in Texas. Running roughly southeast from Hardisty, it would cross the southeast corner of Saskatchewan and then six U.S. states, to a point on the southern border of Nebraska. At that point it would turn south, traversing Kansas, Oklahoma, and Texas and eventually reaching Houston and Port Arthur.

Proponents say the project would contribute to U.S. oil security and job creation, as well as moving mounting oil production to market. The crude now goes chiefly to refineries in the U.S. Midwest, which are projected by 2015 to reach the limit of their capacity to process the heavy oil.

Opponents express concern about what they call significant current air- and water-quality degradation from the extraction of tar-sands crude as well as possible water-contamination risks associated with potential leaks or ruptures during pipeline operation. They note, for example, that the proposed pipeline route crosses the Missouri and Yellowstone Rivers and sections of the huge Ogallala Aquifer that underlies South Dakota, Nebraska, and Kansas.

TransCanada officials said the company would likely apply for a new permit to build the pipeline along a different route.

The pipe is also coated or wrapped to protect it from rust, other chemical attack, and electrical current flow, using materials ranging from fusion-bond epoxy, coal tar, plastics, and tapes to shrink sleeves and even concrete. The protective coating, while sometimes applied at the installation site (fig. 8–5), is commonly applied at special facilities soon after manufacture.

Most pipelines are buried three to six feet below the ground. After welding and coating operations are completed, cranes lift the pipe and lower it into the trench (fig. 8–6). In some situations, the line may be installed on special supports that raise it above ground level, to prevent environmental damage (as was done for some segments of the Trans-Alaska Pipeline System).

Fig. 8–5. Applying spiral-wrap tape to a pipe (*Source:* Miesner and Leffler)

Fig. 8–6. Lowering the pipeline into position (*Source:* Miesner and Leffler)

Other components

A variety of components work together to keep a pipeline operating efficiently and safely. These include:

- Valves (and their actuators) that control or cut off oil flow in response to operational changes, maintenance needs, or emergencies

- Pumps of various designs and capacities that maintain pressure levels to keep the oil flowing (analogous to a compressor in a natural gas pipeline)

- Electric motors, engines, and turbines (running on liquid fuels or natural gas) that provide the power needed to run pumps (fig. 8–7)

- Meters that accurately measure the amount of oil being moved or delivered

- Other instruments, sensors, and computer-linked components that provide real-time information about the condition of all sections of a pipeline

Fig. 8–7. Centrifugal pump driven by an electric motor (*Source:* Miesner and Leffler)

Supervisory control and data acquisition (SCADA) systems are used to control a wide range of large-scale industrial and utility systems, including oil and gas pipelines. In simplified terms, a pipeline SCADA system integrates many of the components described above—plus communications components, visual display units, databases, software, computers, and more—to provide second-by-second information about the condition and operation of a pipeline. Operators can use the SCADA system to optimize pipeline network operation and to take remedial action should the need arise.

Inspection and maintenance

An oil pipeline requires comprehensive inspection and maintenance. This ensures that contractual delivery commitments are met, that the line's useful economic life is maximized, and that it operates safely and efficiently to meet the needs and address the concerns of all stakeholders (fig. 8–8).

One major indicator of potential problems is the leakage of oil or gas. A range of governmental and industry organizations track and report releases from U.S. pipelines. These include the U.S. Department of Transportation (specifically, its Pipeline and Hazardous Materials

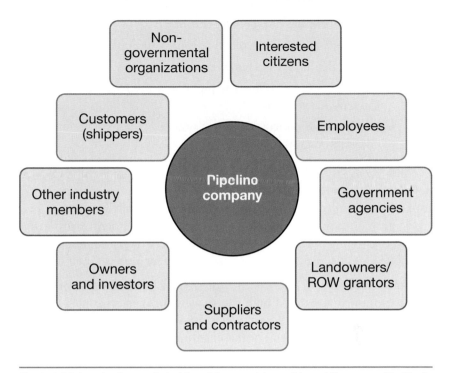

Fig. 8–8. Major pipeline stakeholders (*Source:* Adapted from Miesner and Leffler)

Safety Administration [PHMSA]), the Association of Oil Pipe Lines, the American Society of Mechanical Engineers, API, and the Interstate Natural Gas Association of America. Their reports indicate that the following are the leading causes of pipeline releases:

- Third-party damage to pipelines and equipment from excavation and agricultural activity
- Corrosion (internal and external) and stress corrosion cracking
- Mechanical failures, including manufacturing flaws and construction flaws
- Natural hazards (e.g., ground motion, weather, and erosion due to water flow)

To address these problems, pipeline operators employ a range of strategies, including:

- Compliance with all applicable standards
- Quality control of pipeline fabrication and construction operations

- Careful mapping of pipeline location
- Public education about pipeline routes and operation
- Use of telephone call centers to aid excavators
- Monitoring or patrolling the pipeline route for signs of intrusion or leakage
- Installation of corrosion control systems (e.g., cathodic protection)

The methods and equipment used to find potential problems before they create failures warrant elaboration. Chief among these are sophisticated internal line inspection (ILI) devices, called *smart pigs*, that travel through a pipeline looking for metal loss, wall deformation, and cracks (fig. 8–9). Smart pigs employ advanced technologies, such as ultrasound and magnetic-field measurement, to accomplish their mission.

Another approach is the electrical survey, based on the principle that corrosion is associated with the movement of electrons (a flow of electrical current). Technicians measure the difference in voltage between the pipe and the adjacent soil or the magnitude of current flow between two points on the pipeline.

A third technique, called *direct assessment* (DA), is used in places where ILI tools cannot be employed. DA uses statistical analysis to identify the most likely problem locations along a pipeline. At those points, workers then excavate and examine the condition of the pipe and coating, including measurement of pipe wall thickness.

A final important element is the formal assessment of risk by pipeline operators. In simple terms, risk is calculated by multiplying the probability that an event will happen by the quantitative estimation of the negative consequences that would result from that event. For example, the probability of a major oil pipeline rupture might be very small, say 0.5%. However, if a pipeline were to rupture in a highly populated area, the consequences could include millions of dollars in property damage and business disruption, as well as loss of lives.

An increasing number of U.S. pipeline operators integrate their risk assessments with the kinds of programs described above to create a pipeline *integrity management plan* (IMP). The PHMSA and industry groups have defined what constitutes an acceptable IMP, and a range of vendors offer tools and software that can help pipeline companies create such a plan.

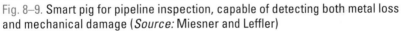

Fig. 8–9. Smart pig for pipeline inspection, capable of detecting both metal loss and mechanical damage (*Source:* Miesner and Leffler)

Pipeline operators also plan and execute a range of programs for the repair and maintenance of pipelines. These include:

- Repair of a pipe's outer surface or coating
- Pipe replacement or reinforcement
- Reduction of stress-concentration points on a pipe
- Burial of a pipeline more deeply for added protection
- Maintenance of cathodic protection systems
- Clearing of pipeline right-of-way (to keep signs and markers visible and to make it easier to see evidence of small leaks or third-party intrusions).

Pigs that are not as "smart" as those described above—but still quite capable—are also used for internal pipeline maintenance. They are run through the line periodically, to keep it free of paraffin, water, sediment, and other debris. This improves product flow and reduces the likelihood of internal corrosion.

Offshore Pipelines

Special technology and techniques are needed to bring crude oil to refineries and other onshore facilities from offshore fields. The challenge is particularly daunting when the production wells are in deep water, typically defined as more than 1,000 feet below the ocean surface.

The *Deepwater Horizon* platform (owned by Transocean and leased to BP), which suffered a catastrophic blowout in April 2010, was connected to wells about 5,000 feet below the surface of the Gulf of Mexico, about 250 miles southeast of Houston. Shell's *Perdido* platform began producing oil and gas in March 2010 from a network of 35 wells at a water depth of about 8,000 feet.

No attempt is made here to describe in detail the complex equipment and advanced skills needed to install a subsea pipeline and its associated components. However, the following sections offer a brief summary.

From wellhead to platform

Several types of pipe and related equipment work together to gather oil or gas from subsea wellheads and bring it to the production platform above (fig. 8–10). These components must be strong enough to handle both the crushing pressure encountered at the seafloor—about 2,700 psi at a depth of 6,000 feet—as well as the internal pressure needed to lift oil to sea level. Piping is commonly made of steel, with wall thicknesses ranging from 0.75 to 2 inches.

Two types of piping are connected to the wellhead. A *jumper*—steel or flexible composite pipe typically 6–12 inches in diameter—carries hydrocarbons to a manifold (see below) if the distance is about 100 feet or less. For longer distances (up to tens of miles), the connecting pipe is called a *flowline*.

The *manifold* is a larger-diameter section of steel pipe to which jumpers from several wellheads can be connected. It has just a single outlet, to which a flowpipe is connected.

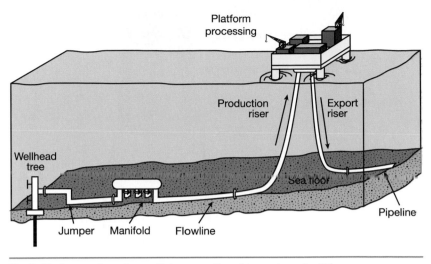

Fig. 8–10. Connections from offshore wellhead to onshore facilities (*Source:* Adapted from Miesner and Leffler)

Lines and manifolds on or close to the seafloor are typically insulated to counteract the effects of water temperatures as low as 30°F. Such temperatures can cause waxy paraffins to build up on the inside walls of an oil pipeline or can trigger formation of hydrate in a line carrying gas or oil. Chemical additives can inhibit paraffin formation, and methanol or glycol can inhibit hydrate formation. (Hydrate is a crystalline material, similar in appearance to a snowball, in which methane molecules are trapped in a lattice of water molecules.)

Steel or composite flowlines, which can be tens of miles long, carry oil or gas to a seafloor connection with another kind of pipe, called a *production riser.* The production riser carries collected hydrocarbons upward to the production platform, sometimes aided by pumps on the seafloor. (A riser can be connected directly to a wellhead if production from only a single well is to be carried upward.) Risers are commonly made of steel, but more flexible types are fabricated by wrapping an inner steel core with alternating layers of steel wires and thermoplastic (fig. 8–11).

Depending on the size of the field being tapped, the water depth, and the location of the production platform, risers can be quite long. For example, the five risers on the *Perdido* platform are each about two miles long, gathering oil and gas from 35 wells drilled into three separate fields within a nine-mile radius in the Gulf of Mexico.

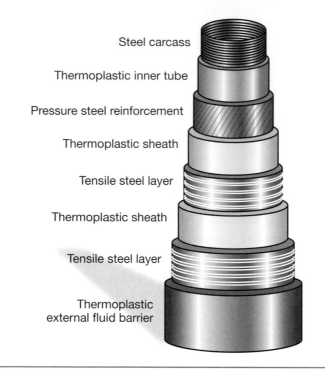

Steel carcass

Thermoplastic inner tube

Pressure steel reinforcement

Thermoplastic sheath

Tensile steel layer

Thermoplastic sheath

Tensile steel layer

Thermoplastic
external fluid barrier

Fig. 8-11. Construction of flexible pipe (*Source:* Miesner and Leffler)

From platform to shore

After crude oil undergoes initial cleanup aboard the production platform, it is either off-loaded to a shuttle tanker for the trip to shore or pumped down through another riser (called an *export riser*) to enter a subsea pipeline. This section focuses on the transport of oil by subsea pipeline.

Laying miles of pipe in water hundreds or thousands of feet deep poses significant technical challenges. The first step is to carefully plan the line's subsea path, evaluating the sea-bottom terrain. The goal is to identify the most cost-effective route that will permit efficient pipeline operation while also minimizing mechanical stress on the installed pipe, as well as any ecological damage.

Once the route is set, pipeline assembly begins. One common method is to weld together and coat pipe sections—typically 40–240 feet long and 16–42 inches in diameter, with walls at least 1 inch thick—at a shore-based facility. Long sections of pipe are then fitted with flotation devices, towed to the installation site, and lowered to the seafloor (fig. 8–12).

Depending on water depth, divers or ROVs connect all the sections together.

Another approach is to weld and coat the pipe sections aboard specially designed pipe-laying vessels that can be as long as a football field. The connected sections are then gradually eased into the water from the stern of the slow-moving vessel as it follows the installation route (fig. 8–13).

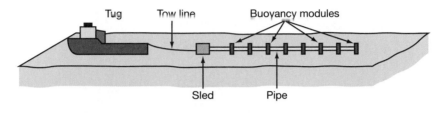
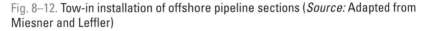

Fig. 8–12. Tow-in installation of offshore pipeline sections (*Source:* Adapted from Miesner and Leffler)

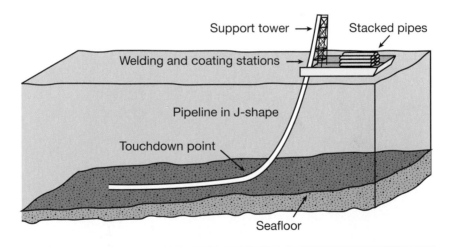

Fig. 8–13. J-lay process of offshore pipeline installation (*Source:* Adapted from Leffler, Pattarozzi, and Sterling)

A third method, suitable only for laying pipe 6–18 inches in diameter, is to weld together and coat the pipe sections on land and then wind the finished product onto very large reels. (More than 30,000 feet of 6-inch pipe can fit on one such reel.) Specially equipped barges carry the reels to the installation site and slowly unwind and lower the pipe into position (fig. 8–14).

Fig. 8–14. Reel barge laying pipeline offshore (*Source:* Miesner and Leffler)

Subsea pipelines can be quite extensive. For example, the Cameron Highway Oil Pipeline System is a 380-mile, 24- and 30-inch oil pipeline that extends along the Outer Continental Shelf of the Gulf of Mexico. It has the capacity to deliver more than 600,000 b/d of oil from several major deepwater fields to the Texas coast, connecting directly with three refineries and four terminals.

Oil Tanker Ships

Crude oil is also transported around the world on seagoing tanker ships (fig. 8–15)—a mode of delivery that is second only to a pipeline in cost-effectiveness. The International Maritime Organization reported a total of 2,105 crude oil tankers in the global fleet in 2008.[1]

These vessels moved about 1.72 billion tons of oil (13.8 billion barrels) to global markets in 2009, according to the U.N. Conference on Trade and Development.[2] This shipment volume was down about 3.4% from 2008 because of the global recession that began in 2008, reducing global oil demand.

Fig. 8–15. Supertanker *AbQaiq*. The vessel is en route to port, to take on an estimated two million barrels of crude oil produced in Iraq. (*Source:* U.S. Navy photograph [2003])

Some 4,954 *oil product* tankers are operated separately, carrying refined products from refineries to markets and terminals around the world in 2008. More information about transport of oil products is presented, along with a discussion of refining, in chapter 11.

The first commercially successful crude oil tankers were developed in the 1870s and 1880s, but World War I was the real impetus for development of the kinds of vessels now used around the world. Decades later, during World War II, the most popular tanker design was the T2. Just over 500 feet long, it carried about 16,500 tons of oil.

Crude oil tankers are typically classified by *deadweight metric tonnage* (DWT)—the displacement of a fully loaded vessel minus its weight when completely empty (with no cargo, fuel, ballast, crew, passengers, water, or provisions). A vessel's cargo-carrying capacity is about 95% of its DWT. One DWT of capacity is equivalent to about 7.5 barrels of oil (though different grades of crude have different densities).

Until 1956, tankers had to be small enough to navigate the Suez Canal. However, when Egypt closed the canal in July of that year, shipowners realized that much larger vessels—able to travel around the Cape of Good Hope, instead of through the canal—could also offer significant transport efficiencies.

The first tanker with a capacity of more than 100,000 DWT was built in 1958. Since then, significant advances have been made in technology and design, as well as size.

The largest tanker ever built was the *Seawise Giant,* fabricated in 1979, with a capacity of 564,763 DWT. Just over 1,500 feet long, with 46 tanks and almost 340,000 square feet of deck area, the ship was said to be too large to pass through the English Channel. The ship had several owners and name changes before being scrapped in 2009.

Crude tankers in use today are required to have double steel hulls so that if the outside hull is damaged, oil will still be contained within the internal hull. Some vessels also have double bottoms for the same reason. Since 2010, no single-hull tanker vessel of 5,000 DWT or larger has been allowed to operate in U.S. waters unless it has a double bottom or double sides, under the terms of the federal Oil Pollution Act of 1990. The act was passed following the March 1989 oil spill from the *Exxon Valdez* tanker in Alaska. The International Maritime Organization, part of the United Nations, enacted a similar double-hull restriction, effective January 1, 2011.

Today's crude oil tankers are also identified by shorthand names related to their size (table 8–3). The term *supertanker* is used for very large crude carriers (VLCCs) and ultralarge crude carriers (ULCCs), which can carry a cargo of more than 2 million barrels.

Table 8–3: Common oceangoing crude oil tanker types

Designation	DWT	Comments
Super Handymax	50,000	
Ultra Handymax	55,000	
Panamax	60,000–80,000	Largest tanker able to pass through the locks of the Panama Canal
Aframax	80,000–120,000	
Suezmax	120,000–180,000	Largest tanker able to pass through the Suez Canal (which has no locks)
VLCC	200,000–320,000	
ULCC	320,000–560,000	

Source: Downey

As of 2010, the four largest supertankers in the world were designated as the *TI* class. Called the *TI Asia, TI Africa, TI Europe,* and *TI Oceania,* the four sister ships were built in 2002–2003 in South Korea, each of approximately 441,500 DWT. Their owners converted the first two to service as stationary FSOs for service in 2009 offshore Qatar.

Crude oil tankers of small or moderate size can dock at a pier designed to pump the oil into or out of the vessel. However, supertankers are typically so large that they cannot do so. Instead, they must take on and off-load their cargoes at offshore facilities or even transfer their cargoes at their port of destination to smaller tankers or barges.

Barges are also used to move oil between onshore storage terminals. In 2010, some 92 million barrels of crude were transported by barge in the United States, according to EIA.

Railcars and Tank Trucks

Two final oil transportation modes worth mention are the tank truck and the railcar. EIA reported that crude shipments from domestic producers to U.S. refineries by tanker truck totaled approximately 145 million barrels in 2010. For railcars, the corresponding figure was roughly 9 million barrels.

Often, oil is moved by truck from small producing wells to regional gathering tanks. This stored oil is then moved farther, by railcar or tank truck, to refineries or main pipelines.

Oil Storage

Crude oil is typically stored in aboveground tanks or in underground caverns. Depending on market conditions and supply/demand dynamics, it is also occasionally stored in tanker ships, as described earlier.

Crude oil storage tanks are typically large cylindrical structures built of carbon steel (fig. 8–16). They are usually painted a light color to reflect the rays of the sun, thereby minimizing evaporation. In some designs, the tank has a roof that floats on the surface of the stored oil. Exposed to the elements or contained under a separate peaked roof, the floating roof prevents evaporation and reduces oil contact with oxygen and moisture. A facility with several storage tanks is called a *tank farm*.

The procedures used to monitor storage-tank volume, oil quality, and related parameters for relatively small tanks were described earlier in this chapter. For large storage tanks, more sophisticated, automated equipment is used.

Fig. 8–16. Oil storage tanks (*Source:* Miesner and Leffler)

Salt caverns are also used to store crude oil (as well as natural gas and propane). The caverns are created within underground salt domes or beds by pumping water into the formation to dissolve the salt. The brine is then pumped out to leave a hollow cavern. Although the salt can be dissolved away by water, it forms a barrier that is impervious to hydrocarbon liquids and gases. A generic illustration of a salt cavern used for oil storage is shown in figure 8–17.

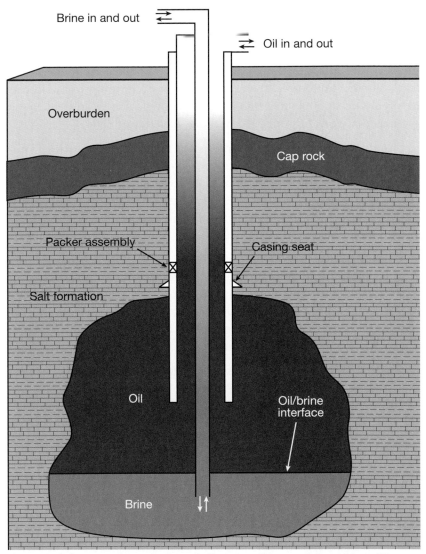

Fig. 8–17. Salt cavern for underground oil storage (*Source:* Adapted from Miesner and Leffler)

A notable example of salt-cavern storage of oil is the U.S. Strategic Petroleum Reserve (SPR). The U.S. and other SPRs are described in detail in the next section.

SPRs

IEA was formed in 1974, in reaction to the embargo imposed on Western nations in 1973 by Arab oil producers. The aim of IEA is to coordinate the response of developed nations to restrictions in oil supply and to gather and disseminate information about global energy markets. One early recommendation of the IEA was that countries with significant levels of oil consumption create stockpiles of oil to mitigate the impact of supply disruptions. In response, the U.S. federal government began construction of its SPR in 1975 and completed it in 1984.

The U.S. SPR consists of four sites, on the Gulf Coast, whose exact locations are not disclosed for security reasons. A total of 60 salt caverns (each typically 2,000 feet high and 200 feet in diameter) have been created at those four sites, capable of holding a combined total of about 725 million barrels of crude oil (sufficient to meet U.S. demand for about two months).

The salt at the bottom of a cavern is about three degrees warmer than at the top, and this differential creates a natural convective flow that keeps the oil slowly circulating. Oil is removed by pumping water into the bottom of the cavern. Oil floats on water, rising to the top of the cavern to be drawn off. It is estimated that it would take 100 days to completely withdraw all of the oil from the U.S. SPR.

Crude has been withdrawn several times from the U.S. SPR:
- In 1985, as part of a test sale
- In 1991, early in the first Persian Gulf War
- In the mid-1990s, to help reduce the budget deficit
- In 2000, during a heating-oil shortage
- In 2005, following Hurricane Katrina
- In mid-2011 (in concert with similar action by the other 27 member nations of the IEA), in response to reduced oil output created by unrest in the Middle East in general and Libya in particular

A number of other nations also have (or are developing) significant strategic crude oil stockpiles. According to EIA, approximately 4.1 billion barrels of oil are held in strategic reserves around the world (including in the United States). Of this total, some 1.4 billion is government controlled, with the remainder held by private industry.

China completed the first phase of construction of a government-controlled SPR in 2008 and expects to have about 500 million barrels of oil in the system by 2016. Japan has about 324 million barrels in such reserves, and India is building a reserve of about 37 million barrels.

Outside the United States, oil is also stored in rock formations. Caverns in the rock must be mined using heavy equipment. Such caverns can be less expensive than aboveground tank storage but are more costly to build than salt-dome caverns.

References

1 International Maritime Organization Maritime Knowledge Centre. October 2009. International Shipping and World Trade Facts and Figures. London. IMO is a specialty agency of the United Nations.

2 U.N. Conference on Trade and Development. 2010. Review of Maritime Transport 2010. Geneva, Switzerland. Published annually by the United Nations Conference on Trade and Development.

Transporting Natural Gas by Pipeline

This chapter and the next focus on the movement of natural gas from the wellhead to the marketplace. After a brief overview of gas transportation, subsequent sections of this chapter "follow the flow" to describe in more detail large-diameter gas transmission pipelines, the role of gas storage, small-diameter local distribution systems, and actions taken to ensure pipeline integrity. While this chapter focuses on pipeline transport, chapter 10 gives specific attention to the transformation of gas to liquid form (liquefied natural gas, LNG) for maritime shipment between countries.

Overview of Gas Transportation

A first look at the transportation of natural gas can be confusing because of the large number of stakeholders involved (fig. 9–1). To simplify the situation, look first at the path taken by natural gas as it moves from the point of production at a wellhead to its entry point into a pipeline (fig. 9–2).

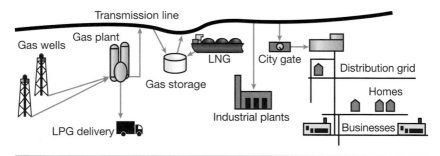

Fig. 9–1. The natural gas value chain (*Source:* Adapted from Miesner and Leffler)

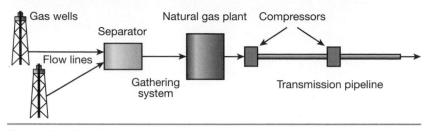

Fig. 9–2. Gas flow from wellhead to transmission pipeline

Natural gas produced from an underground formation first moves through flowlines to one or more separation units, to remove unwanted constituents including water, solids (e.g., sand), and crude oil. The raw gas typically will still contain some water vapor, as well as minor amounts of natural gas liquids (NGLs)—ethane, propane, butanes and pentane. Some nonhydrocarbon gases also may be present, including carbon dioxide, helium, hydrogen sulfide, and nitrogen. The raw gas then flows into a gathering system, which is a network of pipes that carry the gas from the separators to a gas plant, where production volumes can be consolidated for more economical subsequent processing.

At the gas plant, the raw gas is primarily upgraded to meet pipeline quality standards for specific gravity, heating value, and water content (fig. 9–3). At many gas plants, NGLs also are removed, for two reasons: first, they have economic value; second, if left in the gas stream, they can (*a*) affect the Btu content of the natural gas and (*b*) condense into liquid form within the pipe and impede gas flow. Even gas produced by heating LNG (the subject of more detailed discussion in chap. 10) is often processed to remove NGLs. More details about gas processing were presented earlier (see chap. 7).

After gas processing, the natural gas is introduced to a large-diameter transmission pipeline to start its journey to market. Compressors are used to push the gas into the pipeline at the entry point, as well as to maintain sufficient pressure along the pipeline route. In 2011, just over 300,000 miles of transmission lines were operating in the United States, according to EIA, carrying gas for hundreds or thousands of miles on intrastate and interstate routes.

The gas flows to *local distribution companies* (LDCs, also called *gas utilities*) that take control of the gas at critical facilities called *city gate stations*. Using equipment at or close to these stations, the LDCs add an odorant (to aid in leak detection), adjust gas pressure, and then introduce the gas into their distribution networks (fig. 9–4).

Fig. 9–3. Karsto natural gas processing plant, west coast of Norway (*Source: Oil & Gas Journal*)

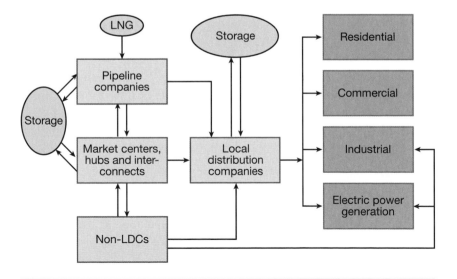

Fig. 9–4. Gas flow from LDCs and other entities to customers

As described in detail starting on page 173, supply mains carry gas to areas where large groups of customers (or large single customers, e.g., factories) are located. At those points, the gas flows into pipes of somewhat smaller diameter, called *feeder mains*. Regulators at various points in the network control gas pressure to ensure reliable delivery at proper pressure levels.

On the last leg of the journey, the gas flows through lines of still smaller diameter, called *service lines* (or *services*). Meters measure the amount of gas used by individual customers.

Description of the processes of planning and building pipelines or distribution systems is beyond the scope of the present text. However, the following sections and chapters provide details on the operation of transmission and distribution pipeline systems.

Transmission Systems

Movement of natural gas by pipeline is a very efficient mode of energy transportation that has been used around the world for many decades. Although this section focuses on the U.S. gas transmission system, it also includes some basic information about gas pipeline systems in other countries.

U.S. gas transmission

High-pressure transport of natural gas in the United States began in 1891, when Indiana Gas and Oil Company laid wrought iron pipe from an Indiana gas field to Chicago, about 120 miles away. Early in the 20th century, large amounts of natural gas were found in Texas, Oklahoma, and Louisiana, and several pipelines were built to carry it to local markets.

It was not until the 1920s, however, that long-distance pipeline transport of gas become economical, based on the development of several important technologies. These included a method for manufacturing seamless steel pipe, the introduction of oxyacetylene (and later, electric arc) welding to join steel pipe into long sections, and the development of instruments for measuring large volumes of gas.

Before 1925, the longest transmission pipelines were only about 300 miles long. However, by 1931, more than a dozen major lines were in service, carrying gas from fields in Texas and Louisiana to cities such as Denver; St. Louis; Kansas City, Missouri; and Chicago. After a pause in construction during the 1930s and 1940s, pipeline installation boomed after World War II, and by 1950, gas pipelines eclipsed those used to carry oil.

Today's U.S. natural gas pipeline network is a highly integrated grid that can move natural gas to and from nearly any location in the Lower 48 states. As of mid-2011, that grid consisted of:

- More than 210 natural gas pipeline systems
- 300,000 miles of interstate and intrastate transmission pipelines (see table 9–1)
- More than 1,400 compressor stations
- More than 11,000 delivery points, 5,000 receipt points, and 1,400 interconnection points for gas transfer
- 24 hubs or market centers that provide additional interconnections
- 400 underground natural gas storage facilities
- 49 locations for the import or export of natural gas by pipeline
- 8 LNG import facilities and 100 LNG peak-shaving facilities

Interstate pipeline companies (table 9–2) are regulated by the Federal Energy Regulatory Commission (FERC). In compliance with major regulatory reforms enacted between 1985 and 1993, interstate pipeline companies have now separated their sales and marketing functions from their transportation operations and provide open access to their pipeline for transportation of gas owned by others. (Open-access rules also apply to underground storage facilities owned by the pipelines.)

Table 9–1. U.S. interstate gas pipeline mileage, 2000–2009

Year	Miles		
	Gas[a]	Oil	Total
2000	186,151	152,823	338,974
2002	190,899	149,619	340,518
2004	190,117	142,200	332,317
2006	189,012	140,407	329,419
2008	192,384	146,822	339,206
2009	192,637	148,622	341,295

Source: FERC annual reports (Form 6 for oil pipelines, Form 2 for gas pipelines). (Adapted from *Oil & Gas Journal*, Nov. 1, 2010, p. 103.)
[a] Includes only FERC-defined major pipelines (transmission mileage).

Table 9–2. Top 10 U.S. interstate gas pipeline companies by mileage, 2009

Company[a]	Transmission mileage
1. Northern Natural Gas	15,028
2. Tennessee Gas Pipeline	14,113
3. El Paso Natural Gas	10,235
4. Columbia Gas Transmission	9,794
5. ANR Pipeline	9,579
6. Transcontinental Gas Pipe Line	9,362
7. Texas Eastern Transmission	9,314
8. Natural Gas Pipeline Company of America	9,312
9. Southern Natural Gas	7,563
10. Gulf South Pipeline	6,565
Total:	100,865

Source: FERC annual reports (Form 2 for natural gas companies), Dec. 31, 2009. (Adapted from *Oil & Gas Journal*, Nov. 1, 2010, p. 103.)
[a] All classified as major companies by FERC.

Kinder Morgan, El Paso deal

In mid–October 2011, Kinder Morgan, based in Houston, announced a definitive agreement to buy El Paso for $20.7 billion. The acquisition will create the largest natural gas pipeline network in the United States, extending some 67,000 miles. Other pipelines for moving petroleum products and carbon dioxide will bring the total length of the company's pipeline system to approximately 80,000 miles.

Under open access, a local distribution company, a large industrial customer, or a group of customers can buy gas directly from a producer or a marketer and contract separately for its transportation by pipeline. Intrastate pipelines are primarily regulated by the states in which they operate. However, they are also subject to certain regulations and the enforcement authority of the FERC, as well as the federal pipeline safety regulations of the U.S. Department of Transportation. Major interstate pipelines are typically 20–42 inches in diameter and constructed of high-strength steel to allow operation at pressures of 200–1,500 psi.

A transmission pipeline system is often referred to as a *mainline* or a *trunk line* if it covers a long distance, uses chiefly large-diameter pipe, and runs directly from a major gas supply source to a market area or an LDC, with few laterals or branches. In comparison, a grid-type transmission system is characterized by a large number of laterals from the main pipeline, creating a network of integrated points for gas delivery and receipt or pipeline interconnection.

When gas transmission piping is fabricated, an internal coating of fusion-bond epoxy may be applied to prevent internal corrosion during shipment and storage, to facilitate internal inspection prior to assembly for welding, and to improve flow efficiency because of the smooth surface created.

In general, by federal rule, a buried transmission line must have a minimum soil cover of 30 inches (in some cases, 36 inches). Also, the line must be installed with at least 12 inches of clearance from any other facilities.

In the United States, a comprehensive federal pipeline safety bill was approved by the Congress in mid-December 2011 and signed into law by President Obama on January 3, 2012. It not only reauthorizes the 2002 federal pipeline safety law but also raises integrity management requirements to new levels. The legislation was driven by several high-profile pipeline accidents in 2010, including a September 9 gas explosion in San Bruno, California (near San Francisco), that killed eight people and destroyed 37 homes.

Basic pipeline equipment

Building and operating a pipeline is a major financial investment, so companies work hard to keep the gas flowing at maximum capacity as much as possible to recoup that investment. They use sophisticated, computer-based control systems to monitor pipeline operation and to schedule movement of gas into and out of storage as needed to minimize peaks and valleys in demand.

Gas enters the pipeline at a point known as an *origination station*. Here, multiple compressors are used to introduce gas from multiple gathering lines into a transmission line (fig. 9–5).

Fig. 9–5. Origination gas compressor station (*Source:* Miesner and Leffler; photograph courtesy of Williams Gas Pipeline)

Because of the friction that occurs when a fluid flows within a pipeline, gas pressure drops with distance and must be boosted. This is achieved by building additional *compressor stations* every 50–100 miles. At these stations (as at the origination station), the gas first passes through separators designed to remove any free liquids or dirt particles from the gas before compression. Any liquids removed are collected and stored for sale or disposal.

Engine/compressor systems. Compressors are the heart of the system, providing the push to keep gas moving through the arteries of the pipeline. They are driven by large engines or motors (prime movers), each of which can generate several thousand horsepower.

There are three basic types of engine/compressor systems. In the first type, a turbine engine (similar to those used on jet aircraft) drives a centrifugal compressor; the centrifugal compressor operates like a fan inside a case, pushing the gas molecules closer together to increase pressure as the fan spins. The second type uses a large electric motor to power a centrifugal compressor (fig. 9–6). The third combines a reciprocating (piston) engine—similar to an automobile engine but much larger—with a reciprocating compressor, whose pistons compress the natural gas (fig. 9–7). Turbine and reciprocating engines typically run on natural gas, drawn from the pipeline.

When a pipeline is designed, the desired gas pressure, pipe diameter, pipe wall thickness, compressor type, and compressor-station spacing are all taken into consideration, to ensure the ability to transport a specified quantity of gas per day at minimum cost. As of early 2011, installed U.S. gas pipeline power totaled about 40 million horsepower.

A variety of technologies have been developed to minimize the release of combustion emissions from turbines and reciprocating engines. (Electric motors, of course, release no emissions at the point of use and are quieter than gas-fueled engines, making them a logical choice for use in populated areas.)

Fig. 9–6. Electric motor driving a centrifugal pump (*Source:* Miesner and Leffler)

Fig. 9–7. Reciprocating engine driving a reciprocating compressor (*Source:* Miesner and Leffler)

Meters. Careful and continual measurement of the amount of gas flowing through a pipeline is essential for both business and safety reasons. Gas producers need accurate records of how much gas they put into a pipeline, how much is delivered to intermediate parties at various transfer points along the pipeline route, and how much is removed at the delivery end. Customers also need accurate measurements to determine payment amounts. Further, pipeline operators, as well as the general public, need to know as soon as possible if gas is leaking from a pipeline.

Measurement of gas volumes is achieved using a variety of sophisticated and carefully designed meter technologies. Meter types include:

- *Positive displacement:* Measuring flow in discrete segments, similar to people moving through a turnstile, and then adding up the segments

- *Turbine:* Measuring flow rate by monitoring the speed of rotation of a rotor suspended in the flow stream (fig. 9–8)

- *Orifice:* Forcing the gas to flow through a plate containing a carefully machined hole of known size and then using Bernoulli's principle (relating mass flow, velocity, and pressure difference upstream and downstream from the orifice) to calculate flow rate

- *Ultrasonic:* Sending high-frequency pulses of sound energy across the pipe diameter, then using data on flow velocity, pipe cross-sectional area, and gas density to calculate flow rate

- *Coriolis:* Using the rather esoteric principle of the Coriolis force (based on angular momentum) to measure the deflection of a vibrating tube and, thereby, the mass of natural gas moving through the meter

Fig. 9–8. Turbine meter (*Source:* Miesner and Leffler)

Meters are subject to wear and deterioration with use. They must also be carefully calibrated on a regular basis to ensure accuracy and to enable the correction of raw measurements to standard pressure and temperature conditions.

Other components. Many other components are employed in pipeline systems. These include many kinds of specialized instrumentation, signal transmitters, sensors, valves, actuators, and computers as well as more mundane fittings, flanges, scrubbers, and strainers. All of these contribute to the efficient, safe, and reliable delivery of gas by pipeline. Detailed description of these components are beyond the scope of this book (See app. B for other reading suggestions.)

Gas storage

Storage is an integral part of any system for delivery of large volumes of natural gas. The primary reason is that demand for gas can vary— with the time of day, the season, and the operating schedules of large-volume customers, such as factories. In addition, natural gas production capacity can change over time, and storage can be used as a way to absorb increased volumes or as a gas source if production drops.

Pipeline companies use a system called *operating storage* to balance short-term swings in demand. This entails an operator's moving gas into storage at night and extracting it early in the day, on a daily basis, year-round.

Gas can be stored in a variety of ways; the following are the most common:

- Under pressure in depleted oil or natural gas reservoirs
- In aquifers (water-bearing sedimentary rock formations; see fig. 9–9)
- In underground salt caverns (as described in chap. 8)

In 2009, EIA reported that most (331 of 409) underground storage facilities in the United States were depleted reservoirs. Of the remainder, 43 were aquifers, and 35 were salt caverns.

During the past 20 years, the number of salt cavern storage sites has grown significantly for two reasons: first, the capability of such facilities to achieve rapid cycling (inventory turnover); second, their ability to respond to daily, even hourly, variations in customer demand.

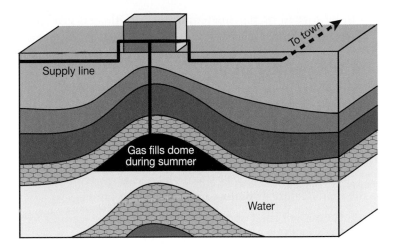

Fig. 9–9. Aquifer gas storage field (*Source:* Adapted from Busby, R.)

As noted previously (in chap. 8), the large majority of salt cavern storage facilities have been developed in salt-dome formations located in the Gulf Coast states. Gas stored in that region takes 4–12 days to reach markets in the Northeast.

In discussing underground storage, it is important to define three terms:

- *Total gas storage capacity:* The maximum volume of gas that can be stored in an underground storage facility based on its design
- *Base gas* (or *cushion gas*): The volume of gas that must stay in a storage reservoir to maintain adequate pressure and deliverability rates throughout the withdrawal season
- *Working gas:* The volume of gas in the reservoir above the level of base gas (in other words, the amount available to the marketplace)

In February 2011, EIA reported a total of 6,029 bcf of natural gas held in underground storage in the Lower 48 (by all operators of such facilities). Of this total, 4,305 bcf was base gas, leaving 1,724 bcf as working gas.

Gas can also be stored in aboveground steel storage tanks, either pressurized or partially pressurized and refrigerated.

In a final important source of short-term storage called *line pack*, more gas is pushed into a pipeline than is withdrawn, being careful not to exceed the maximum allowable operating pressure of the pipeline.

Storage can be owned by pipeline companies, LDCs, gas producers, dedicated gas storage companies, and even some gas trading companies.

Inspection and maintenance

Much of the information presented in chapter 8 regarding inspection and maintenance of oil pipelines applies as well to natural gas pipelines. The gas pipeline industry devotes considerable effort to a range of actions designed to maintain pipeline integrity and ensure safe operation, including:

- Adherence to rigorous quality and physical-property standards in the manufacture, handling, and assembly of piping

- Careful route selection and installation of pipeline sections to minimize possible damage from flooding, soil movement, or other natural hazards

- Application of coatings to protect external pipe walls from mechanical damage and corrosion

- Use of cathodic protection and related electrical-survey systems to control corrosion and thereby minimize metal loss from pipeline walls due to galvanic or microbial action

- Conduct of regular airborne and on-foot patrols of the pipeline right-of-way to detect any incursions, as well as signs of equipment damage or gas leakage

- Performance of hydrostatic tests on new pipeline sections (and, occasionally, on existing sections) to check for leaks by filling the line with water, raising the pressure to the expected operating level, and holding that pressure for a predetermined time while monitoring for leakage

- Education of the public and construction professionals (including use of maps, telephone call centers, and route markers) on the location and depth of pipelines

- Periodic cleaning of interior pipeline walls with pigs—pushed along by gas pressure—that remove water, sediment, paraffin, and other debris to help prevent corrosion and reduce resistance to gas flow

- Use of more-sophisticated smart pigs (ILI tools) that move through a pipeline to nondestructively examine the condition of the pipe wall and pinpoint any areas of concern (fig. 9–10)

- Occasional direct assessment (i.e., excavation and visual inspection) of pipe sections that ILI tools cannot traverse or that statistical analysis suggests may be prone to coating or pipe-wall deterioration
- Compliance with regulations established by government agencies regarding pipeline operation, maintenance, and repair
- Timely repair or replacement of any pipeline section or component as needed.

Fig. 9–10. Multi-diameter ILI tool prepared for launch (*Source: Oil & Gas Journal*)

Global gas transmission

More than 50 natural gas transmission pipelines are operating in countries outside North America (table 9–3). Another dozen are planned or under construction.

Table 9–3. Selected major gas pipelines outside North America

Pipeline location/name	Nation(s)[a]	Maximum capacity, bcf/year
Africa:		
West African	Nigeria, Benin, Togo, Ghana	177
Asia:		
Arab	Egypt, Jordan, Syria, Lebanon, Turkey	364
Central Asia-Center	Turkmenistan, Uzbekistan, Kazakhstan, Russia	3,178
Central Asia-China	Turkmenistan, Uzbekistan, Kazakhstan, China	1,413
Dolphin	Qatar, UAE, Oman	706–1,059
West-East	Shaanxi Province, Beijing	565 (2 lines)
South Caucasus (or BTE)	Azerbaijan (Caspian coast), Georgia, Turkey	706
West-East	Xinjiang Province to Shanghai, China	600
Europe:		
Balgzand Bacton Line (BBL)	Netherlands, UK	565
Blue Stream	Russia, Turkey	565
Europipe I and II	Norway, Germany	1,483
Franpipe	North Sea (Norway), France	692
Gasdotto Algeria Sardegna Italia (GALSI)	Algeria, Italy	353
Greenstream	Libya, Italy	389
Jamal-Gas-Anbindungsleitung (JAGAL)	Germany	848
Langeled	Norway, UK	900
Maghreb-Europe	Algeria, Morocco, Spain, Portugal	424
Mittel-Europaische-Gasleitung (MEGAL)	Germany	777
South Wales (Milford Haven)	Wales	742
Soyuz	Russia, Western Europe	1,130
Trans Austria (TAG)	Austria	1,677
Trans-Mediterranean (TransMed)	Algeria, Tunisia, Italy	1,067
Yamal-Europe	Russia (Western Siberia), Belarus, Poland	1,165
Zeepipe	North Sea (Norway), Belgium	2,366 (3 lines)

Source: Wikipedia.

[a] Origin point listed first.

Distribution Systems

As of early 2011, about 71 million customers in the United States relied on natural gas for a range of energy services, and about two-thirds of the gas they used was delivered by some 1,500 LDCs. There are two basic types of LDCs: private firms owned by investors, and public gas systems owned by local municipalities. According to EIA, LDCs deliver 99% of the gas used in the residential sector and 98% of the gas used in the commercial sector. In the industrial and electric power generation sectors, LDCs play a lesser role, delivering only about 45% and 2%, respectively, of the gas used; the remainder used in these latter two sectors comes directly from pipelines or other non-LDC entities (e.g., gas marketers).

City gate station

As noted earlier in this chapter, the city gate station is the point at which natural gas moves from the transmission pipeline system into the local distribution network. At or near the gate station, water and impurities are removed, the volume of the incoming gas is measured for billing and gas-control purposes, the gas pressure is reduced, and an odorant (typically mercaptan) is added to aid in leak detection.

Near the city gate station, the LDC also may add supplemental supplies, to meet spikes in demand caused by weather fluctuations or to adjust the heating value of the gas. Supplemental supply can include propane, butane, and LNG.

Piping network

Gas typically flows into a city gate station at a pressure of 1,000–1,500 psi. This pressure is reduced in a series of steps as it flows through the distribution system to the facilities of various kinds of customers.

Five types of piping are used in the typical gas distribution system:
- Supply mains carry gas at 100–200 psi from city gate stations to regulator stations that reduce the pressure and move the gas into feeder mains. (Supply mains may deliver gas at high pressure to some customers, e.g., large industrial facilities.)
- Feeder mains typically deliver gas at 50–60 psi—downstream from a regulator station—to distribution mains. (As is the case for supply mains, feeder mains may connect directly to some customers).

- Other regulators at or near the origination point of distribution mains step down the pressure further, primarily for supply to residential, smaller commercial, and industrial services. Distribution mains are often buried under city streets.
- Service lines deliver gas from distribution mains to the meters of various customers.
- Fuel lines—pipes inside a building ("beyond the meter")— bring gas to appliances. These lines are the property and the responsibility of the building owner.

Many distribution systems consist of overlapping networks of mains operated at different pressure levels. In figure 9–11, for example, note the service line to a large industrial customer directly from a supply main. Such customers require specially designed meter and regulator sets to handle the large gas volumes and higher delivery pressures they need.

Small commercial and residential customers served from a high-pressure supply main require an additional regulator to reduce the supply line pressure from the service line. That regulator will generally reduce the pressure to 60 psi or less.

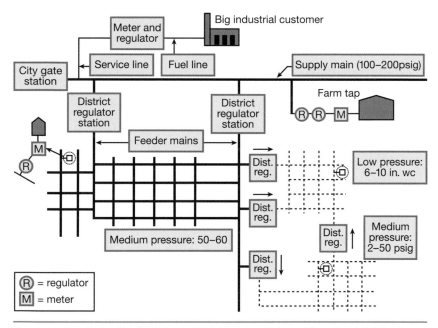

Fig. 9–11. Local gas distribution network (*Source:* Adapted from Busby, R.)

Feeder mains deliver gas into low-, medium-, and high-pressure sections of the distribution system. Newer systems typically operate at a source pressure of 60 psi and require a regulator on each service line to further reduce the pressure to levels required for appliances.

If necessary, other district regulator stations drop feeder-main gas pressure further for use in older parts of the distribution system. The individual services connected to mains in these older distribution systems also require regulators, to reduce the pressure to levels necessary for proper appliance operation.

Sections of many older LDCs operate at very low pressure—in the range of 6–10 inches of water column (0.22–0.36 psi). Household appliances are generally designed to operate near this pressure level, so individual service regulators are not needed. In fact, many modern appliances have built-in regulators to ensure proper performance.

Piping materials

Pipes used most commonly in a gas distribution system are made of steel, cast iron, ductile iron, plastic, or copper, as detailed below:

- Steel pipe has strength and flexibility but will corrode quickly if it is not coated and cathodically protected. Pipe joints on steel lines are typically welded.

- Cast iron pipe does not corrode as quickly as other metals but is more brittle than steel or plastic. It may break when subjected to forces such as ground settlement, frost, or traffic vibration.

- Ductile iron pipe is more flexible than cast iron and therefore is more resistant to breaking. It does, however, corrode in a manner similar to cast iron.

- Plastic pipe does not corrode but is softer and much easier to scratch or cut than steel. A metal wire must be buried along with plastic pipe to permit workers to confirm its location at a later time if necessary by using a metal detector.

- Copper tubing is flexible (allowing use in confined spaces), resistant to corrosion, easy to connect, and available in long lengths. The American Society for Testing and Materials specifies the types of copper tubing suitable for use in gas distribution systems.

Supply, feeder, and distribution mains are made of cast iron or ductile iron. Service lines can be fabricated from plastic, steel, or copper. Copper is the traditional material of choice for fuel lines inside a building, although flexible corrugated stainless steel tubing (coated in plastic) is finding increased use.

Mechanical compression is the only practical method for joining cast iron and ductile iron pipes. In this process, a rubber gasket is compressed against the pipe ends by metal flanges to create a gas-tight seal. Plastic and steel pipe also can be mechanically joined if welding or fusion (described below) are too difficult or present a safety hazard (e.g., if potentially explosive gases are present). Sections of plastic pipe (and fittings used with them) are typically joined with adhesives or by a process called *heat fusion*—melting the plastic and applying pressure to fuse the pieces together.

In general, distribution mains must be buried with at least 24 inches of cover. However, exceptions can be made if a state or municipality either allows installation in a common trench (along with other utility lines) or provides damage prevention by other means. At least 18 inches of cover are required for service lines buried under streets, and 12 inches is the minimum for lines on private property.

Minimizing gas leakage

Local distribution companies, like pipeline companies, work hard to ensure worker and public safety and reliable operation—in particular, by minimizing gas leakage. Such efforts also improve LDC economics (by reducing gas loss) and reduce the impact of atmospheric release of methane, a potent greenhouse gas.

Because mechanical damage is one of the leading causes of pipeline failure, considerable effort is devoted to the prevention of contact with the piping system and related equipment. This includes public education programs, as well as one-call systems (telephone hotlines that provide pipeline-location data to anyone planning an excavation project).

Leak surveys are widely used to ensure that all gas piping is inspected regularly. Gas can be detected both above- and belowground by using sensitive instruments carried by hand or mounted on vehicles. Survey crews also watch for another indicator of leakage—namely, the change in color of vegetation, from green to brown or yellow, due to the drying effect of escaping gas.

A third element in leak detection is *odorization*. Because natural gas has virtually no smell, addition of an odorant (generally at the city gate station) is required by federal regulation. The odorant (commonly mercaptan) gives natural gas what consumers recognize as a rotten-egg smell. Sufficient odorant is added to ensure that gas is detectable at a level of about 1% (or less) in air. When a leak is suspected, various tools are used to pinpoint it, excavate the suspect pipe section in the least disruptive way, repair the pipe, and restore the excavation site cost-effectively.

Controlling corrosion

Corrosion is a chemical reaction that returns a refined metal to its natural state as an ore. Contributing factors include heat, moisture, seawater, soil contaminants (e.g., salts and fertilizers), time, mechanical stress, and temperature.

Corrosion of buried pipe—the leading cause of leaks from buried gas mains—occurs when a small current of electricity flows from the pipe into the ground, causing deterioration of the pipe material. As a pipe corrodes, its wall thickness is reduced, creating the potential for gas loss or outright failure. Replacement of corroded pipe imposes additional maintenance costs as well.

Management of corrosion is required under Part 192 of the U.S. *Code of Federal Regulations*, and a range of methods and devices are used to do so. Although coatings and insulation can help, another common approach is *cathodic protection*. In simple terms, cathodic protection manages the flow of electrical currents in the walls of a pipeline in a way that minimizes or eliminates corrosion damage.

Corrosion on the outside wall of a pipeline is called *external* or *atmospheric corrosion* and is accelerated in areas of humidity, pollution, and hot or fluctuating temperatures. It can be caused by improper application or poor bonding of pipe coating, improper pipe maintenance, contact between incompatible metals, coating damage, or a localized corrosion cell created by abnormal conditions.

Corrosion of a pipe's inner wall, or *internal corrosion*, results from the presence of water, bacteria, or chemicals. Just as with transmission pipelines, routine pigging to clean the pipe wall helps to reduce internal corrosion.

Peak-shaving

As noted previously, gas can be retrieved from storage facilities to manage variations in gas demand. However, when withdrawals from storage are insufficient to meet peak customer requirements, LDCs may use a technique called *peak-shaving*.

In most cases, peak needs are met by vaporizing LNG or LP gas (such as propane) and injecting it into the distribution system supply (fig. 9–12). Short-term line packing can also be used to meet an anticipated surge in demand. During the gas shortages of the 1970s, many LDCs installed synthetic natural gas (SNG) plants that used naphtha as feedstock to supplement curtailed natural gas supplies. SNG is expensive to produce, however, and these plants are not currently used for baseload service.

Whatever supplemental fuel is used, the goal is to cut (shave) as much of the difference as possible between the total maximum user requirements (on a peak day or shorter period) and the baseload customer requirements (the normal or average) daily usage. Doing so evens out the supply/demand curve for a pipeline, thereby reducing costs and increasing pipeline utilization. Also, by storing gas when demand (and price) is low and then putting the gas back into the pipeline when demand (and price) is high, a local gas utility can operate more efficiently.

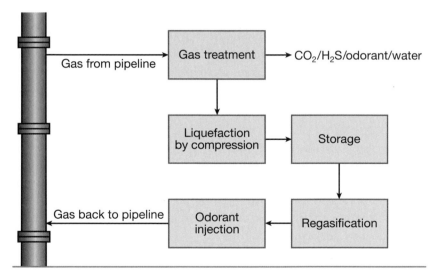

Fig. 9–12. Peak-shaving plant using LNG (*Source:* Adapted from Kidnay, Parrish, and McCartney)

10 Transporting Natural Gas as LNG

An increasingly important development in the movement of natural gas—particularly between nations—has been the evolution, over the past several decades, of the LNG industry. Transforming natural gas into a cryogenic (ultracold) liquid increases its energy density dramatically, permitting movement by ship, barge, or tank truck instead of by pipeline. The same cryogenic technique is also used to expedite shipment of industrial gases such as nitrogen, oxygen, and hydrogen.

Background

The three major steps in the LNG value chain (fig. 10–1) are the liquefaction process, the shipment of the LNG (most often by ships of special design) to an import terminal, and the transformation of the LNG back into gaseous form (regasification) for introduction into a pipeline that will carry it to market. Interim storage of the LNG is also part of industry operations at both liquefaction plants and regasification plants.

When natural gas (at atmospheric pressure and 60°F) is cooled below its boiling point (–263°F), its volume can be reduced by a factor of more than 600, and it can be stored and shipped without pressurization in liquid form. These are distinct economic benefits for gas shipment.

Liquefaction can be particularly valuable in bringing to market *stranded gas*—that is, gas that is not located close to users and that would require significant investment in infrastructure to transport it to market. Examples include gas resources found far offshore and in very deep water, as well as gas resources in far northern Arctic regions. LNG also allows countries to import natural gas cost-effectively from other nations around the world.

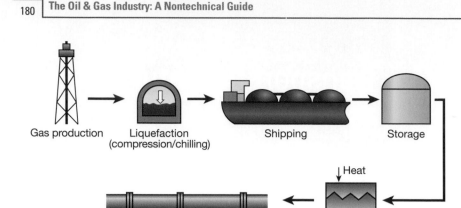

Gas production   Liquefaction   Shipping   Storage
(compression/chilling)

↓ Heat

Pipeline transport   Regasification

Fig. 10–1. LNG value chain (*Source:* Adapted from Tusiani and Shearer)

LNG will remain an ultracold liquid with no need for a cooling system because of the naturally occurring process called *boil-off*. In this process, some natural gas vapors evaporate from the LNG, and this evaporation draws heat away from the liquid, keeping it cold. Effective thermal insulation will maintain the low temperature of the LNG as long as the boil-off gas is removed from the storage vessel.

The first use of LNG for peak-shaving by a local gas distribution company occurred in the late 1930s in the United States. Maritime transport began in 1959, with the shipment of an LNG cargo from Louisiana to the United Kingdom. In 1964, the first global commercial-scale chain for LNG production and delivery went into operation, liquefying natural gas in Algeria and sending it to import terminals in France and the United Kingdom.

The 1960s and 1970s saw a jump in construction of both liquefaction and import facilities, followed by a slowdown in the 1980s (except in Asia). By the early 1990s, demand for natural gas in the Western hemisphere was catching up with supply, and several new LNG projects were launched, becoming operational in the late 1990s.

By 2010, some 24 liquefaction facilities and 80 import (regasification) facilities were in operation around the world, served by a tanker fleet of about 330 vessels. An additional 25 liquefaction plants were being planned or under construction (including 10 in North America alone), as were an additional 32 regasification plants.

As of mid-2011, there were 11 U.S. facilities (as well as another in Peñuelas, Puerto Rico) capable of importing and regasifying LNG. They are located in:

- Everett, Massachusetts
- Cove Point, Maryland
- Elba Island, Georgia
- Lake Charles, Louisiana
- Gulf of Mexico, offshore Louisiana
- Offshore Boston (2)
- Freeport, Texas
- Sabine, Louisiana
- Hackberry, Louisiana
- Sabine Pass, Texas

A facility in Kenai, Alaska—from which LNG had been exported to Japan and China since 1969—was mothballed in late 2011, though owner ConocoPhillips said in December that it expected to resume limited operation in mid-2012. On the basis of increasing U.S. gas supplies (chiefly from exploitation of gas-bearing shales), proposals were made in late 2010 to build or retrofit one facility in Texas and another in Louisiana for the export of LNG from the United States to world markets. In addition, more than 100 LDCs were operating smaller LNG regasification facilities to help meet peak gas demand.

A variety of hurdles must be overcome in the development and implementation of a large-scale LNG project, in addition to the obvious engineering challenges inherent in handling a high-energy-content cryogenic material. These include:

- Assurance of sufficient gas reserves over the project life (typically 20–30 years)
- Access to large amounts of capital
- Gaining long-term commitments from buyers of LNG
- Arrangements for use of tanker vessels
- Negotiation of multiple and complex commercial issues such as contracts, ownership rights, insurance, revenue sharing among partners, and financing
- Compliance with governmental regulations regarding safety, environmental impacts, security, and economic policies (e.g., tariffs) for both land-based facilities and vessels

The Liquefaction Process

Natural gas from various formations and regions will differ in composition. It is primarily methane, with small amounts of ethane, propane, butane, and pentane. It may also contain condensates (light oils) and other constituents.

Because LNG is bought and sold on the basis of its heating value, constituents that affect heating value must be removed. For example, ethane, propane, butane, or pentane will raise the heating value of methane, while nitrogen and carbon dioxide will reduce it. In addition, unless carbon dioxide, water, hydrogen sulfide, and mercury are extracted, they will freeze at low temperature, forming crystals that could plug filters, heat exchangers, and other equipment. Condensates are also removed.

The process of removing carbon dioxide or hydrogen sulfide is called *sweetening*. Removal of water or water vapor is called *dehydration*.

After it has been cleaned, the pressurized incoming gas undergoes a refrigeration process by which it is cooled to about –240°F (fig. 10–2). A final temperature reduction to –260°F is then carried out by flashing the gas (i.e., rapidly reducing its pressure by passing it through a valve) to just above atmospheric pressure.

The system of equipment used to liquefy LNG is called a *train*. The typical liquefaction facility has a half-dozen or more trains set up in parallel, and larger plants can produce up to eight million tons per year of LNG. A range of proprietary systems have been developed for liquefaction, but the following are the primary equipment in a train:

• Compressors, typically driven by natural gas–fueled turbine engines

• Heat exchangers, in which the heat from the incoming compressed gas is transferred to refrigerant gases (e.g., propane or ethylene) and then either to the atmosphere or to a stream of water

• Flash valve, to achieve final cooling

The resultant LNG is a clear, colorless, odorless liquid with half the density of water. It is pumped into special insulated tanks of various designs for storage before being loaded onto ships for transport to market.

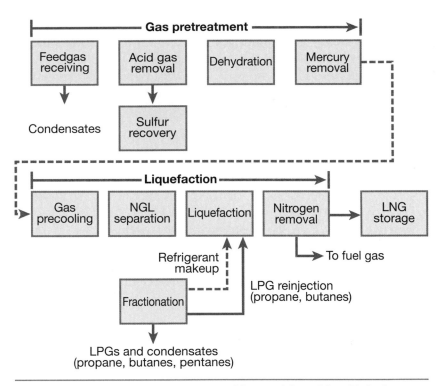

Fig. 10–2. Natural gas liquefaction process (*Source:* Adapted from Tusiani and Shearer)

LNG Storage Tanks

It is typical to install storage-tank capacity equivalent to about twice the cargo capacity of the largest LNG tanker ship expected at a liquefaction plant. Aboveground tanks are constructed more often, although underground or partially buried tanks are used in some densely populated areas (e.g., Tokyo Bay) for aesthetic reasons. All LNG storage tanks have double walls, and tank pressure is maintained at just above atmospheric pressure (15–20 psi).

The three most common onshore LNG storage tanks are the single-containment, double-containment, and full-containment designs. The first of these is the simplest and least expensive. It consists of a freestanding, open-top inner tank of 9% nickel steel, plus a carbon steel outer tank, with several feet of thermal insulation between the two. A steel roof contains any boil-off vapor and supports a suspended ceiling that insulates the top of the inner tank. The term *single containment* means that the outer tank is not designed to contain any LNG that escapes from the inner tank; instead, the escaped LNG would flow into a secondary containment system, typically an area around the tank that is ringed by dikes.

Double containment is similar to single except that the outer tank (made of reinforced concrete and strengthened by a wall of earth or rock) can contain any LNG that leaks from the inner tank. A steel roof protects the inner tank but cannot hold any vapor released from it.

The third tank design is called *full containment* (fig. 10–3). It adds a concrete roof to the double-containment system's concrete outer walls. This roof can hold any vapor produced by a breach in the inner 9% nickel steel tank. Although this design is the most expensive of the three—about 50% more costly than single containment—it allows the closest spacing between tanks and equipment.

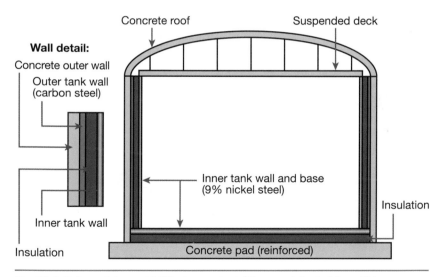

Fig. 10–3. Full-containment LNG storage tank (*Source:* Adapted from Kidnay, Parrish, and McCartney)

A fourth design, called a *membrane storage tank*, is seldom used. In this design, a flexible stainless steel membrane is supported by a layer of insulation that is mounted on an outer prestressed concrete wall. The outer concrete wall and roof can contain any leak of liquid or vapor from the primary membrane. Membrane tanks are generally seen as less durable than other designs.

At most LNG facilities, LNG is loaded from storage onto tanker ships moored at a jetty with loading arms incorporating specially designed couplings (fig. 10–4). A cryogenic piping system carries the LNG from the tanks to the jetty. It can take 24 hours to moor a tanker, connect the loading system, and slowly cool the LNG cargo tanks prior to loading. A typical jetty can handle about 215 loadings per year (about 13 million tons of LNG) servicing larger tankers.

Fig. 10–4. Darwin LNG plant, at Wickham Point, Australia (*Source: Oil & Gas Journal*)

Offshore Liquefaction Concepts

As demand for LNG rose from 2000 to 2005, various parties proposed innovative concepts for fixed or floating systems targeted at the exploitation of small or remote offshore reserves of stranded gas. One design, intended for use in shallow water, is the *gravity-based structure* (GBS), similar to technology used to build some offshore oil and gas production facilities. The GBS provides an artificial island for LNG production and storage, using large prefabricated concrete structures that are flooded and sunk to rest on the seafloor.

Another approach involves the use of a specially designed FPSO moored in deeper water to take gas from a subsea gas field below it and convert it into LNG onboard. From such a floating LNG (FLNG) vessel, the LNG is later transferred to conventional LNG carriers that occasionally tie up to the FLNG vessel.

In May 2011, Royal Dutch Shell announced plans to build a FLNG vessel for use 125 miles off Australia's northwest coast (fig. 10–5). The company described it as the biggest floating human-made object ever built—longer than four football fields and more massive than an aircraft carrier. It would convert 110,000 boe of gas from the Prelude field, where production is slated to start by 2017.

Fig. 10–5. Proposed Shell FPSO (FLNG vessel) for liquefaction offshore Australia (*Source: Oil & Gas Journal*)

LNG Tanker Ships

The cargo tanks (containment system) on an LNG tanker must provide a gas-tight seal to prevent mixing of gas vapor with air; insulate the LNG from heat, to minimize boil-off; and prevent the steel hull of the ship from becoming very cold, which could make it brittle. Two major designs have been widely adopted by the industry: the self-supporting independent tank and the membrane tank.

The self-supporting independent tank is a rigid and heavy structure of either spherical (Moss or Kvaerner Moss) design or box shape. In the Moss design (fig. 10–6), the LNG is stored in several large aluminum tanks, the top half of which protrude above the deck.

Fig. 10–6. Statoil's *Arctic Princess*, an LNG tanker with Moss cargo tanks (*Source:* Statoil)

In the membrane design, the tank walls are built into the ship after the hull has been constructed and are located belowdecks (fig. 10–7). A primary membrane (of nickel steel alloy or stainless steel) is designed to be in contact with the LNG, while a secondary membrane provides protection in case the primary membrane leaks. The secondary membrane, although typically of the same material as the primary membrane, may sometimes be made of aluminum sandwiched between glass cloth. Polyurethane foam or perlite insulation is installed between the membranes, as well as between the secondary membrane and the ship's hull.

Fig. 10–7. Qatar Gas Transport's *Mozah*, an LNG tanker with membrane cargo tanks (*Source: Oil & Gas Journal*)

Both tank designs are built to withstand the weight of the LNG. In turn, the tanker's hull is designed to support the rigid tank structure.

The membrane tank is box-shaped and is fabricated from light, flexible metal. Rigid, load-bearing insulation is installed between this metal membrane and the tanker's hull. The insulation not only allows the transfer of loads from the LNG tank to the vessel's hull but also lets the membrane surface contract and expand as its temperature changes.

A novel tank geometry known as *structural prism* was developed about 2005 and is in limited use. This design is based largely on the membrane concept but incorporates a high, peaked top and a long, narrow, liquid free surface. It aims to reduce the structural loads that can be imposed when LNG sloshes around inside a cargo tank.

Regardless of the containment system, care must be taken to avoid imposing a thermal shock on the LNG tank when a cargo is loaded. In traditional practice, a small volume of LNG (called *LNG heel*) is left in the tank throughout a vessel's trading period. This reduces the time and cost needed to cool the tanks down before every loading and minimizes imposition of multiple thermal cycles (warm-to-cold-to-warm), which can stress the containment system.

Two European shipyards (in France and Spain) can build LNG tankers, but most construction is now done in South Korea and Japan owing to lower labor costs. China is also increasing its ability to construct LNG tankers.

Of some 330 LNG tankers in global service in early 2010 (fig. 10–8), cargo capacity ranged from less than 25,000 cubic meters (883,000 cubic feet) to more than 200,000 cubic meters (7.06 million cubic feet). About 65% of these vessels had a capacity of 125,000–150,000 cubic meters (4.4 million to 5.3 million cubic feet). Construction cost for a tanker in the latter capacity range is approximately $160 million to $175 million.

A tanker carrying 4.7 million cubic feet of LNG will discharge about 60,000 tons of liquid that becomes about 3 bcf of gas when regasified. The largest vessels are about 1,200 feet long and 180 feet wide and can travel at about 20 knots. A typical LNG tanker crew would number about 35 members.

Most LNG vessels are dedicated to a particular project or trade route for the 20-year life of a typical LNG contract. In recent years, however, some ships have been built on a speculative basis for short- or medium-term LNG trading.

LNG SHIPPING FLEET

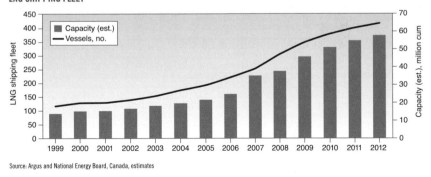

Source: Argus and National Energy Board, Canada, estimates

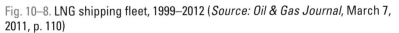

Fig. 10–8. LNG shipping fleet, 1999–2012 (*Source: Oil & Gas Journal,* March 7, 2011, p. 110)

Onshore Regasification Process

Once an LNG tanker completes its voyage, it typically moors at a large marine LNG receiving terminal, often built on the coast or on the shore of a river with easy access to the ocean (fig. 10–9). These facilities receive LNG in large quantities and pump it back into large storage tanks of the kind described earlier to be held until needed. (One exception: The proprietary Bishop Process bypasses the cryogenic storage step. The LNG flows from the tanker through a pipe-in-pipe heat exchanger that uses the heat from circulating seawater to vaporize the LNG immediately. The gas is then pressurized and sent either directly into a pipeline or into an underground salt cavern.)

A marine LNG terminal typically has several storage tanks, each able to hold up to 160,000 cubic meters (3.8 bcf) of natural gas. Four tanks of that volume can hold an amount of LNG about equal to the cargoes of two to four LNG tankers—sufficient to maintain a suitable operating inventory. The sendout (of gaseous natural gas) from an LNG terminal is a function of pipeline capacity, storage and vaporization capacity, and frequency of LNG cargo delivery.

Fig. 10–9. Canaport marine LNG receiving terminal, St. John, New Brunswick (*Source: Oil & Gas Journal*)

At the appropriate time, workers at the terminal convert the LNG back into the gaseous state by a process commonly called *regasification* (fig. 10–10). This process routes the LNG through vaporizers (heat exchangers) in which it is brought into indirect contact with seawater, air, the exhaust from a fuel-burning heater, or heated water. The natural gas may be used on-site at an industrial plant or nearby, in a power plant; alternatively, it may be introduced into a pipeline for delivery to customers in distant markets.

On a smaller scale, the same basic regasification process is utilized by more than 100 LDCs in the United States that use LNG for peak shaving (as described previously). The LNG, delivered by pipeline or in special tanker trucks, is stored in insulated tanks until needed.

Brief mention of a third kind of LNG facility is warranted: the satellite LNG plant. Such plants are mini-regasification facilities, typically owned and operated by industrial customers. Located in areas not served by a pipeline or where pipeline capacity is inadequate, they typically receive LNG by truck or barge and store it in insulated tanks holding up to 10,000 gallons, to be regasified as needed.

Regardless of the size of the regasification facility, odorant is added if necessary. This is the final step before moving the regasified LNG into a pipeline for delivery to customers.

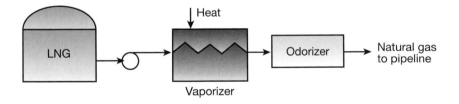

Fig. 10–10. Typical LNG regasification process (*Source:* Adapted from Kidnay, Parrish, and McCartney)

Offshore Regasification Process

In the mid-2000s, other approaches to regasification were moving from concept to commercial application: the conversion of LNG to gaseous form aboard converted or specially designed vessels moored offshore.

One approach involves use of a tanker that contains both LNG tanks and a regasification system. After arrival on station, the vessel connects to a special buoy that is in turn connected to a pipeline on the seabed. The tanker's LNG is regasified onboard and sent down to the pipeline that carries it to shore.

Another is the floating storage and regasification unit (FSRU)—either a custom-built ship or a converted LNG tanker—typically moored permanently at an offshore site. Other loaded LNG tankers tie up to the FSRU and pump their LNG into the FSRU's insulated tanks. Later, the LNG is regasified and sent to shore in the manner described above.

In a third concept, a specially built system is attached to a conventional LNG tanker offshore. This pulls LNG from the tanker and vaporizes it. Then the LNG is sent through a subsea line to an onshore pipeline.

Offshore regasification systems such as these will need to prove their safety, reliability, and economic feasibility in the coming years before being widely adopted.

LNG Safety

A wide range of codes, standards, industry guidelines and governmental regulations spell out the actions necessary for proper construction, careful maintenance and safe operation of LNG facilities and equipment. In the United States, the following federal agencies regulate various aspects of the LNG industry:

- FERC
- DOE
- U.S. Coast Guard
- Department of Transportation
- Department of Homeland Security
- EPA

- Bureau of Ocean Energy Management, Regulation, and Enforcement
- Army Corps of Engineers
- National Oceanic and Atmospheric Administration
- Department of Labor/Occupational Safety and Health Administration

State and local agencies (e.g., fire, police, and environmental departments) also may have requirements regarding construction and operation of LNG terminals. For example, the National Fire Protection Association has developed standard NFPA 59A, which pertains to LNG production, storage, and handling. NFPA 59A incorporates a range of industry standards developed by such professional organizations as the American Society of Civil Engineers, the American Society of Mechanical Engineers, the American Concrete Institute, API, and the American Society for Testing and Materials. The worldwide LNG industry also follows closely the guidance of the International Maritime Organization (an agency of the United Nations) regarding safety, environmental issues, legal issues, and technical operation of LNG tanker vessels.

11 Converting Oil into Products

Crude oil is the starting point for creation of a wide range of products, ranging from gasoline and diesel fuel to roofing tar and heating oil. In turn, petrochemicals made from oil are used to create such products as plastics, cosmetics, insect repellents, and floor wax.

However, raw crude oil (as noted earlier) is a complex mix of hydrocarbon molecules of varying size and structure and is unsuitable for direct use in anything but some industrial boilers. Because of this variable molecular structure, different crude oils can vary widely in appearance and in physical properties. One may be dark brown and as thick as molasses; another may have a green tinge and flow like water; and a third may be black as coal and smell like rotten eggs.

Overview of Refining

The range of processes used to transform raw crude oil into usable products is conducted in a complex industrial facility called a *refinery* (fig. 11–1). Operators of a refinery commonly accept as feedstock a mix of crude oil types (the *crude slate*) from various sources and use a series of processing units to transform this slate into finished products.

In the marketplace, these finished products are often classified as follows:

- *Light distillates*: liquefied petroleum gas (LPG), gasoline, naphtha
- *Middle distillates*: kerosene and related jet engine fuels, diesel fuel
- *Heavy distillates and residuum*: heavy fuel oil, lubricating oils, paraffin wax, asphalt/tar, petroleum coke

Importantly, initial processing takes place at or near the well site, soon after the flow of crude oil begins. Almost every oil reservoir has some gas dissolved in the oil; conversely, almost every gas reservoir holds some oil. For economic reasons, it makes sense to do a bit of cleanup of both hydrocarbons (gas and oil) close to the wellhead (fig. 11–2).

Fig. 11–1. Tesoro's Golden Eagle refinery, Martinez, CA (*Source: Oil & Gas Journal*)

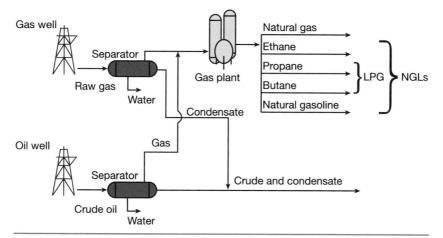

Fig. 11–2. Initial processing of hydrocarbons in the oil patch (*Source:* Adapted from Leffler)

As it comes to the surface and leaves the wellhead, the crude oil stream (containing dissolved gases) is directed into a separator vessel. As the stream enters the vessel, it undergoes a drop in pressure, and the light dissolved gases come out of the oil. (This effect is similar to what happens when a bottle of club soda is opened.) These gases can then be drawn from the top of the vessel for further processing (as described in chap. 7), while the oil remains in the lower part of the separator.

The crude is also very likely to contain some water or water vapor, and that water is drawn from the bottom of the separator. (Recall that oil floats on water.) The oil is then withdrawn, often mixed with condensate (a very light crude oil) from the gas separator, and sent on its way to be refined.

At the refinery, the transformation of crude oil into finished products is generally carried out in four phases: separation/distillation, conversion, enhancement, and blending. These are each covered in the following sections.

Separation/Distillation

In the separation state, the first issue of concern is the removal of salts and any remaining water from the crude oil stream. Significant volumes of water are often produced along with crude oil, coming from the underground formation or resulting from either waterflooding or other efforts to boost production. Much of the produced water is removed close to the wellhead (by the separator, described above, and other processes), but crude entering a refinery typically still contains enough water to require treatment. This water often contains dissolved salts (calcium, sodium, and magnesium chloride compounds) that can damage refinery equipment.

The incoming crude is typically first routed to a vessel called a *desalter*, in which extra water is added and the crude/water solution is intensely mixed and heated to about 250°F. An electric field is then used to separate the crude from the water droplets, and chemicals remove the salts from the water. For removal of the final traces of water, the crude is typically routed to settling tanks or to a centrifugal separation system.

After the removal of water and salt, the crude undergoes a process called *distillation*. As a brief review, consider the use of distillation to purify water that contains some nonvolatile impurities (e.g., potassium or sodium). The water is heated and as the temperature reaches 212°F, it begins to boil. The water molecules turn to vapor, rising above the liquid surface. When that vapor is drawn away from the heated vessel and cooled, it condenses (returns to liquid form), yielding water that is more pure than the original liquid (because only the water molecules vaporized). In the remaining hot water, the concentration of nonvolatile contaminants increases. When that water is boiled away completely, the contaminants are left behind as a solid residue on the walls of the vessel. In this way, distillation separates contaminants from water.

Unlike water, crude oil is not chemically homogeneous. It contains many different hydrocarbon molecules, each having a distinct temperature range over which boiling takes place (fig. 11–3). The smaller the molecule is, the lower its initial boiling point will be.

For example, butanes and lighter petroleum gases will begin to boil off at about 90°F. The gasoline fraction will vaporize between 90°F and 220°F. Around 800°F, even heavy residual hydrocarbons will finally begin to boil.

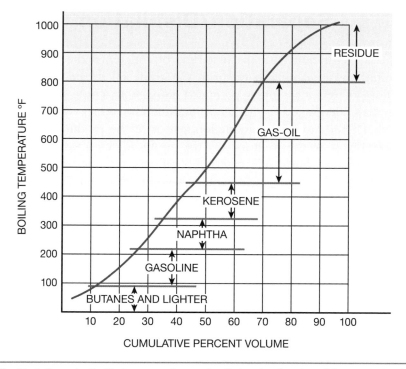

Fig. 11–3. Generic distillation curve for crude oil showing fractions (*Source:* Leffler)

In an oil refinery, *atmospheric distillation* is the simplest form of this process, so named because it is carried out in a tall, cylindrical vessel (a distillation column or tower) operating at atmospheric pressure. This vessel is also called a *fractionating tower* because it separates the raw crude into fractions (or *cuts*) that consist of molecules of more nearly uniform size (compared to the raw crude). Later, these can be processed more easily into finished products.

As one might guess, different grades of crude (which sell at different prices based on their qualities) have different distillation curves (fig. 11–4). In preparation for entry into the distillation tower, crude oil is routed through a furnace in which superheated steam raises its temperature to about 750°F. The hot crude oil is then introduced at the bottom of the distillation tower, flowing onto the lowest tray in a stack of horizontal trays that fill the tower, each separated by one to two feet.

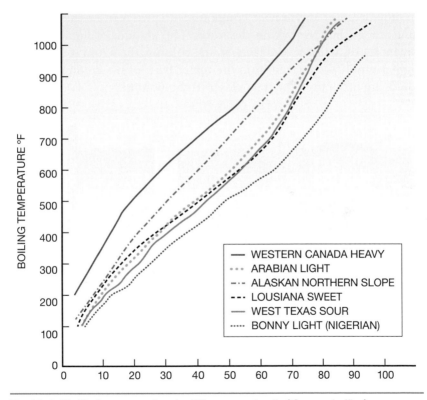

Fig. 11–4. Distillation curves for six different crude oils (*Source:* Leffler)

(Note that the design of the components inside a distillation column can vary widely. Components of various sizes, shapes, and materials can be used, tailored to the type of feed expected and the degree of uniformity required in the fractions produced. The structure and operation of a simple, generic column is described here to illustrate the major aspects of the distillation process.)

A high temperature is maintained at the bottom of the distillation column, and as the hot crude vaporizes, perforations in each tray allow the vapor to rise within the column. As it rises, the vapor cools as it bubbles through the several inches of liquid that eventually builds up in each tray. This heat transfer is enhanced by the use of devices such as *bubble caps* (fig. 11–5). Their geometry forces the vapor bubbles to spend more time in the liquid in each tray, which cools the vapor further. Some hydrocarbons in the vapor condense into liquid form and remain in the tray.

The temperature of the vapor drops, and the lower temperature of the liquid causes any heavier compounds that remain in the vapor to condense as they rise within the tower. The amount of liquid in each tray increases, and at several levels in the column, the liquid is drawn off. Lighter products are taken from the upper parts of the column; heavier liquids are taken from the trays closer to the bottom.

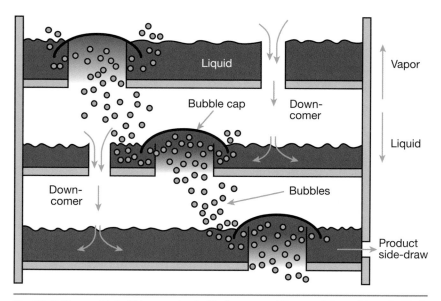

Fig. 11–5. Trays with bubble caps in distillation tower (*Source:* Adapted from Leffler)

An ancillary unit catches any heavy-hydrocarbon vapor that makes it to the top of the column, cools it back to liquid form, and sends it back into one of the lower trays. Conversely, some light hydrocarbons may get mixed into the liquid part of the heated crude at the bottom of the column. This liquid is sent through a reboiler that drives off the lighter hydrocarbons, which can then be reintroduced at some higher level in the tower.

Note that the heavy residuals are typically subjected to a second stage of separation in a *vacuum distillation unit* that operates at a pressure below atmospheric. This decreases its operating temperature and avoids damaging the residuals with excessive heat.

In summary, distillation separates the crude oil into a range of specific hydrocarbon compounds that can be further processed in other parts of the refinery (fig. 11–6). This downstream processing is discussed in the next sections.

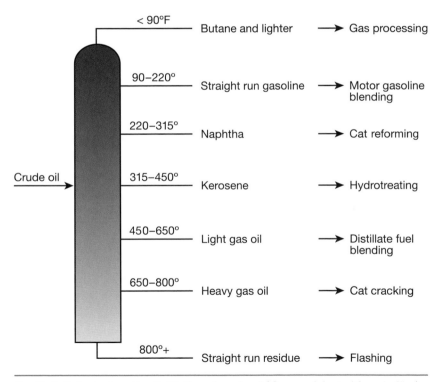

Fig. 11–6. Outputs from the distillation of crude oil (*Source:* Adapted from Leffler)

Conversion

In the refining business, *motor gasoline* and compounds called *middle distillates* are the finished products in highest demand and with the largest profit margins. There is a strong incentive to maximize production of these two product groups by converting as many of the molecules in crude oil as possible into these forms.

Conversion is the broad term applied to processes that crack (break apart), combine, or modify nongasoline and non–middle-distillate hydrocarbons into the two product groups noted above. (Conversion is also sometimes called *upgrading*, because it transforms lesser-value products in ways that increase their value.) Conversion also can be used to produce petrochemicals and to reduce the viscosity of residual fuel oil. Each of the major conversion processes is discussed individually below.

Cracking

Cracking is the breaking of large hydrocarbon molecules into smaller ones. Thermal cracking uses heat to achieve this end, while catalytic cracking uses catalysts to do so. (Recall that catalysts are compounds that promote a chemical reaction but do not take part in it.) In some cases, both approaches also use pressure. Cracking is carried out in tall structures with thick walls, and the resulting stream of hydrocarbons generally must be sent through a distillation column to separate its product fractions.

Thermal cracking. There are several related types of thermal cracking. Those of most importance to oil production are described below.

Steam cracking. One type of thermal cracking, steam cracking makes use of high-temperature steam to do its work (as the name implies). Steam cracking can be used in several ways. In one configuration, it processes heavy residues from the distillation tower (fig. 11–7) to yield lighter products such as gasoline, naphtha, and gas oils.

Steam cracking can also be used to break apart lighter molecules— propane, ethane, naphtha, light fuel oil, and gas oil—to produce petrochemicals. The hydrogen in the high-temperature steam bonds with the cracked molecules to form new ones—primarily high-value ethylene and propylene. (Petrochemicals production is discussed later in this chapter.) When used in this second configuration, the processing unit is sometimes called an *ethylene cracker*. However, it also yields such valuable by-products as benzene, light fuel oil, hydrogen, and methane.

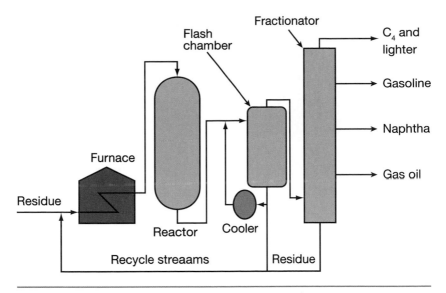

Fig. 11–7. Simplified thermal (steam) cracker schematic (*Source:* Adapted from Leffler)

Visbreaking. A second kind of thermal cracking, visbreaking is used to reduce the viscosity of heavy residual fuel oil. It uses hardware that is simpler and cheaper to build than the hardware used in a steam cracker. The residual fuel it produces is more directly usable in power plant boilers and in marine engines. The visbreaker unit also produces some blendstock for gasoline, naphtha, middle distillates, and bitumen.

Coking. Coking is the final thermal cracking process to be defined here. It converts the residuals from distillation into more-valuable gases, middle distillates, gasoline blendstock, and naphtha, plus a by-product called coke. The coke is a heavy, coal-like material that must be blasted out of the barrel-shaped coker by use of high-pressure water jets. It has value as a fuel for utility boilers and is used to make electrodes for industrial electric melters.

The most common design, the *delayed coker*, has specific design features that slow down the coke formation process (fig. 11–8). The feed to a delayed coker is typically heavy oil and residuals, introduced into large insulated vessels called *coke drums*, where it is allowed to "cook" slowly. Lighter, more-valuable products are drawn off from the top of the drums, and the solid coke (virtually all carbon) builds up at the bottom. Other coker designs have been developed—called *fluid cokers* and *flexicokers*—but not many of these types have been built.

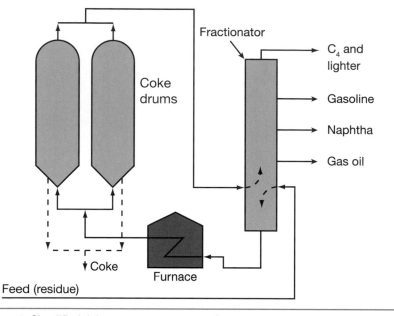

Fig. 11–8. Simplified delayed-coker schematic (*Source:* Adapted from Leffler)

Note that several refinery units are deliberately operated in ways designed to *prevent* the creation of coke during intermediate process steps. Heater tubes in a furnace, for example, can become *coked up* when carbon atoms crack off from a feedstock and adhere to the inside of the tubes, clogging them. Coking problems are prevented by adding steam, by maintaining high velocities in feed streams, and by reducing pressure to reduce process temperatures.

Catalytic cracking. As in thermal cracking, catalytic cracking (also called *cat cracking*) uses heat and pressure, but it also employs a catalyst to either speed up the process or allow it to occur at a lower temperature. A simplified schematic of a catalytic cracking unit is shown in figure 11–9.

In the form of small beads, pellets, or powder, the catalyst is added to the oil and processed along with it. The feed material (heavy residuals from the distillation column) and hot catalyst (about 700°F) are heated further with steam to about 1,000°F, driving the combination upward through a riser. Almost all of the cracking reactions take place in the riser in just a few seconds. Then, in the disengagement vessel, cyclones separate the catalyst and send it to be either disposed of or cleaned (to

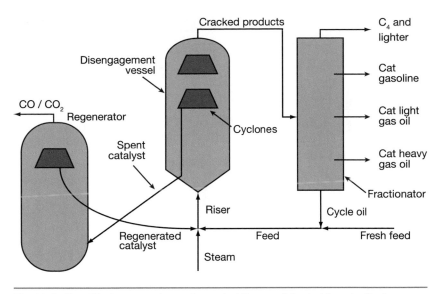

Fig. 11–9. Simplified cat-cracker schematic (*Source:* Adapted from Leffler)

remove the carbon that builds up on its surface) for reuse. The desired cracked products are drawn from the top of the unit.

Catalytic cracking has almost completely replaced thermal cracking because the former produces more gasoline with a higher octane rating as well as more-valuable by-product gases. The two common types of cat crackers are described below.

Fluid catalytic cracker (FCC). There are several proprietary FCC design configurations, but in general terms, all employ a similar mode of feedstock/catalyst interaction: The particles of heated catalyst powder behave like a fluid because they are suspended within a stream of incoming hydrocarbon feedstock (typically heavy gas oil). In this fluidlike state, the catalyst can react more easily with the feedstock, improving the speed and efficiency of the cracking process.

Hydrocracking. Though similar to FCC, hydrocracking uses hydrogen, lower temperatures, higher pressures, and different catalysts (fig. 11–10). It is more expensive and requires a lot of energy to operate. Because it adds hydrogen as it cracks the feedstock (typically heavy vacuum gas oil), it converts virtually all of that feedstock to the desired products (typically high-value gasoline, jet fuel, or middle distillates).

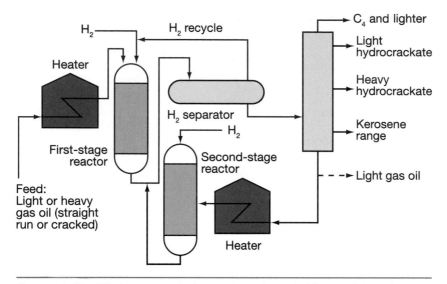

Fig. 11–10. Simplified two-stage hydrocracker schematic (*Source:* Adapted from Leffler)

Combination

The opposite of cracking is combination—the joining together of smaller hydrocarbon molecules to produce larger, more valuable ones. Two common combination methods are polymerization and alkylation.

Polymerization. Polymerization puts together a series of molecules. One common approach uses heat (or lower-temperature heat in combination with a catalyst) to combine light olefin gases, especially propylene and butylene, into a higher-octane gasoline blendstock called *polymerate.* The feedstock for a polymerization unit can come either from a cracking process or from a refinery's light-ends unit or gas plant.

A version of polymerization called *dimerization* links just two molecules—isobutylene and hydrogen—to create isooctane. The latter is an octane-raising gasoline blendstock that can replace MTBE (methyl tertiary butyl ether), a water-soluble octane enhancer. Use of MTBE has been widely discontinued in the United States owing to its negative impact on groundwater quality when it leaks or spills.

Alkylation. Like polymerization, alkylation is used to put together propylene and butylene (from a cracking unit), but it uses isobutane and acidic catalysts to produce the desired higher-octane gasoline blendstock. The process (fig. 11–11) also yields small amounts of propane and normal butane as by-products. Alkylation has largely replaced polymerization (a somewhat older process) in most refineries.

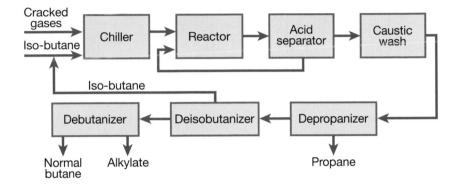

Fig. 11–11. Alkylation plant schematic (*Source:* Adapted from Leffler)

Modification

Refiners also use modification to achieve finished products with desired properties by the rearrangement or alteration of hydrocarbon molecules. Two important modification processes are catalytic reforming and isomerization.

Catalytic reforming. Catalytic reforming (also called *cat reforming*) modifies naphtha into high-octane blendstocks for gasoline—compounds called *reformates*—by using pressure, heat, and a catalyst (typically metals from the platinum group). Fig. 11–12 shows one design—called a semiregenerative (or fixed-bed) cat reformer—in which excess hydrogen levels are maintained in the reactors to minimize deposition of carbon on the catalyst. (The carbon reacts with the hydrogen to form methane and ethane, and this process regenerates the catalyst to some degree.)

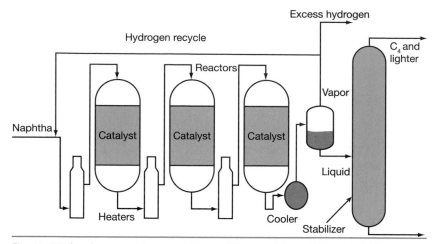

Fig. 11–12. Semiregenerative cat reformer (*Source:* Adapted from Leffler)

Materials called *aromatics* (benzene, toluene, ethylbenzene, and isomers of xylene [BTEX]) can often be separated from reformate to use in petrochemicals or as additional gasoline blendstocks. Hydrogen is a major by-product of reforming, and the naphtha cat reformer is often the primary source of hydrogen for other uses in a refinery.

Isomerization. This is the process used to create isomers. An isomer is a molecule or compound that has the same number of atoms as another but has a different arrangement of those atoms, which produces different physical and chemical properties.

In a refinery, isomerization is used to rearrange the molecules in normal paraffins (butane, pentane, and hexane) to yield isoparaffins (isobutane, isopentane, and isohexane, respectively), which have a higher octane rating. Isobutane is used as a feedstock for alkylation (described earlier), while the other two isomers are used as gasoline blendstocks.

Enhancement

In general terms, enhancement is the use of selected processes to remove unwanted elements or compounds from a hydrocarbon at some stage in the refinery. The most common enhancements are hydroprocessing, amine treating, solvent extraction, and sweetening, discussed in the following sections.

Hydroprocessing

Hydroprocessing uses hydrogen to remove sulfur, nitrogen, nickel, and vanadium from gasoline and middle distillates (kerosene, diesel, jet fuel, and heating oil). As noted earlier, the hydrogen often comes from the refinery's naphtha cat reformer. If more is needed, then hydrogen can be generated by using steam to reform methane. Hydroprocessing is also sometimes called *catalytic hydrotreating* or *catalytic hydrodesulfurization.*

Amine treating

Also sometimes called *amine washing*, this process uses amine solvents to remove hydrogen sulfide, a highly toxic gas, from oil. (An amine is an organic compound derived from ammonia by replacing one or more hydrogen atoms with organic radicals.) The hydrogen sulfide can form during hydroprocessing. Common amine solvents are MEA (monoethanolamine) and DEA (diethanolamine).

Solvent extraction

A third common enhancement process, solvent extraction is also known as *solvent recovery*. At various stages in the refinery system, operators introduce a solvent into a product stream to selectively remove some component. The solvent must have two key characteristics: first, the component must dissolve completely into the solvent; second, it must be possible to easily remove the solvent/component solution from the product stream. Solvents are used in amine treating, BTEX recovery, and removal of waxes and asphalt-like molecules from the residue of a distillation tower.

Sweetening

Sweetening is the process used to neutralize the sulfur compound mercaptan in gasoline and other intermediate and finished products. Mercaptan has a strong, characteristic rotten-egg smell, and sweetening oxidizes it into odorless disulfides that can remain in the fuel. This process is accomplished in a module sometimes called a *merox unit.*

Other enhancement processes are used to produce bitumen, lube oil, and grease. However, their description is beyond the scope of this book.

Blending/Finishing

Most finished petroleum products that end up in the marketplace are the result of blending a range of hydrocarbon molecules from a number of different processes at a refinery. Perhaps the most notable example is gasoline. As it comes from the distillation tower, *straight-run* gasoline has an octane rating too low to permit direct use in modern engines. Various components (blendstocks) are mixed into this gasoline to produce the gasoline sold to consumers.

Gasoline blendstocks

Refiners face a challenge in choosing among blendstocks that have different octane ratings, vapor pressures, and sulfur content. Complicating the picture, blendstocks can interact with each other in ways that further alter these parameters.

The following are the most widely used gasoline blendstocks:
- FCC gasoline (i.e., the output from an FCC)
- Reformate (heavy naphtha, from the distillation tower, that has been catalytically reformed)
- Naphtha
- Alkylate (created by combining propylene and butylene from a cracking unit)
- Isomerate (isopentane and isohexane)

One important gasoline blendstock—ethanol—is added not at the refinery but downstream, typically at the storage depot of a wholesale gasoline distributor. This gives the refiner flexibility in selling its gasoline (i.e., some distributors may want to offer gasoline without ethanol in it). This practice also avoids an operational problem, because ethanol attracts water, which would contaminate gasoline moved by pipeline.

Gasoline additives

Other finishing compounds (distinct from blendstocks) are gasoline additives, mixed into the fuel by refiners and oil companies to boost engine performance or maintain fuel quality. Common additives are detergents, dyes, antifoaming agents, emulsifiers, anti-icing agents, corrosion inhibitors, spark-plug-deposit preventers, and oxidation inhibitors.

Types of Refineries

Classification by complexity

Refineries are often classified into five categories according to their complexity:

- Basic or topping
- Hydroskimming
- Cracking
- Coking
- Full conversion

Basic or topping refineries (used chiefly in China) simply separate the components of crude oil into various products by distillation. They are not capable of more-complex operations, such as reforming or cracking. Their partially finished output is typically sold to other refineries for the processing that will bring them into compliance with basic product-quality specifications.

Hydroskimming refineries conduct the basic separation process done by topping facilities but also have naphtha reformers. These reformers generate hydrogen, which can be "skimmed off" for use in desulfurization units, such as hydrotreaters (fig. 11–13).

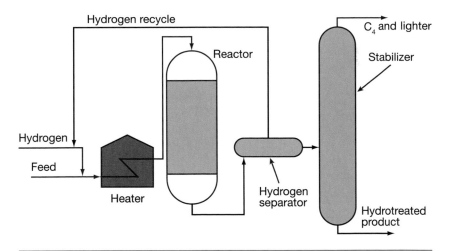

Fig. 11–13. Hydrotreater schematic (*Source:* Adapted from Leffler)

Cracking refineries have catalytic cracking units (e.g., an FCC or a hydrocracker) in addition to modules, such as an alkylation plant, that can increase the yield of high-value products (e.g., gasoline).

Coking refineries have the same capabilities as cracking refineries but also have cokers that can turn heavy vacuum residue into higher-value products.

A full-conversion (or complex) refinery is essentially a petrochemical plant coupled with an oil refinery. Such a facility not only has all the capabilities of a coking refinery but also has a steam cracker. This latter unit is used to produce ethylene and its major by-products (e.g., propylene). Ethylene and propylene are the building blocks for producing most plastics. Refineries have become steadily more complex as reflected in the increasing yield of gasoline and middle distillates since the 1980s.

Classification by capacity

Refineries also are classified by the magnitude of their processing capability. This capacity can be the *charge capacity* (the daily rate of feedstock input) or the *production capacity* (daily rate of final-product output). *Nameplate capacity* is the total amount of product that a refinery is designed to produce in a full calendar year.

Refinery Maintenance Schedules

Because refineries are expensive to build, owners prefer to operate them continuously. However, mechanical equipment wears out, catalysts become depleted, and new equipment must occasionally be installed; therefore, every refinery has a schedule for the performance of routine maintenance, called a *turnaround*. A turnaround can take as little as one day but more often lasts several weeks.

Turnarounds are typically conducted at the same time each year, in the spring and fall, when demand for all petroleum products is relatively low as compared to the winter heating season and the summer driving season. The spring turnaround is generally more extensive than the fall turnaround.

Exact turnaround dates are often published in advance, so that traders and market analysts can factor the maintenance schedules into their supply/demand projections. EIA tracks and reports capacity utilization, a measure of how close to full output U.S. refineries are operating.

Product Pipelines

The variety of outputs from refineries are collectively referred to as *petroleum products*. As noted in chapter 8, a network of pipelines—separate from those that move crude oil—carry these products to intermediaries or customers.

Finished petroleum products

An important component in the transport of petroleum products to customers is the product pipeline, carrying such commodities as gasoline, jet fuel, and diesel fuel from refineries, trading centers, or marine terminals to markets. The delivery points could be the facilities of a product distributor (e.g., a gasoline wholesaler), a trucking company, another trading company, or an export company. Several major product pipelines and their routes are shown in table 11–1.

In many cases, a single kind of product is delivered in large quantity. However, products can also be moved through a pipeline in smaller batches, with multiple products lined up one after the other (fig. 11–14). Often, no physical barrier is needed between the batches; because of the nature of pipeline flow and the physical properties of each product type, very little mixing results.

Table 11–1. Selected major U.S. product pipelines

Name	Route
Colonial Pipeline	Houston to New York
Explorer Pipeline	Gulf Coast to Chicago
Plantation Pipeline	Gulf Coast to Washington, DC, area
Buckeye Pipe Line	New York Harbor to Northeast and Midwest
Williams Pipeline	Tulsa, OK, to Minnesota and Wisconsin
Teppco	Gulf Coast to Chicago

Source: Downey.

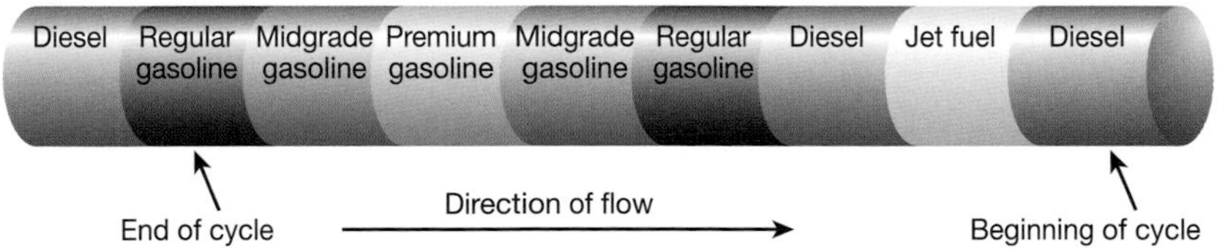

Fig. 11–14. Typical refined-product batch sequence (*Source:* Adapted from Miesner and Leffler)

In some cases, a small degree of mixing can be tolerated—for example, between kerosene and low-sulfur diesel fuel. However, if the two products are not generally compatible (e.g., gasoline and low-sulfur diesel), the *transmix* that is created between the batches must be extracted and reprocessed at the delivery end of the pipeline. If absolutely no mixing or contamination can be tolerated, a sphere or other mechanical device can be inserted between product batches to keep them segregated.

Petrochemicals and LPGs

More information about the production of petrochemicals is presented in a later section. Although no attempt will be made here to delve into this complex subject in detail, it is helpful to consider the characteristics that guide decisions regarding pipeline transport of petrochemicals and LPGs.

Whereas crude and refined products are essentially incompressible liquids, petrochemicals and LPGs are chemically and physically different, requiring shippers and pipeline operators to be much more concerned about temperatures and pressures (including such parameters as vapor pressure, critical-point temperature, and critical-point pressure), as well as product compressibility. The most common of these products consist of molecules with two, three, or four carbon atoms. (This differentiates them from methane—with just one carbon atom—and from gasoline, crude oil, and other complex hydrocarbons that have five or more carbon atoms.) Table 11–2 shows the industry shorthand used to describe petrochemicals and LPGs as C_2, C_3, and C_4 products.

Table 11–2. C_2–C_4 petrochemicals and LPGs

Product	Chemical formula	Classification
Ethylene	C_2H_4	C_2
Ethane	C_2H_6	C_2
Propane	C_3H_8	C_3
Propylene	C_3H_6	C_3
Butanes	C_4H_{10}	C_4
Butadienes	C_4H_6	C_4
Butylenes	C_4H_8	C_4

Source: Leffler

It is preferred to transport C_2 products through pipelines in gaseous, rather than liquid, form. Thus, pipelines carrying ethylene typically operate at temperatures above about 50°F and at pressures between 800 and 2,100 psi. At the lower end of this range, ethylene behaves more like a gas; at the higher end, it behaves more like a liquid. In comparison, ethane pipelines usually operate at temperatures below 90°F and pressures not much above 700 psi.

C_3 products are transported through pipelines as liquids at typical pipeline pressures ranging from 300 to about 1,400 psi. C_4 products also are most often transported in liquid form through pipelines, at operating pressures ranging from 100 to 1,100 psi.

The compressibility of petrochemicals and LPGs requires pipeline operators to use special techniques to measure product flow and to detect leaks. They also must pay close attention to maintaining product quality, particularly for ethylene and polymer-grade polyethylene. For quality-control reasons, petrochemical pipelines are almost always dedicated to transport of a single product, and specialized instruments are used to monitor product quality.

Before a pipeline can be brought into service for petrochemical transport, all moisture, rust, scale, and sludge must be removed from its interior. It also must be purged with nitrogen to remove any oxygen. In addition, care must be taken to control the large drop in temperature that occurs (due to pressure letdown) during initial product introduction. For example, the temperature can drop to –40°F when introducing C_3 products and to –120°F for C_2 products. Pipelines are typically made with carbon steel, which is rated only for temperatures down to –20°F, so the pipeline operator must either maintain some backpressure in the line during product filling or add heat during the input process.

The same caution must be used when removing product from the line to permit repair or maintenance. In many cases, the product can be displaced to a storage facility by using nitrogen. However, if the product cannot be easily handled, then it must be flared or burned as fuel.

Large quantities of petrochemicals and LPGs (typically 100,000 to 1 million barrels) often must be stored under high pressure. This can be done in specially designed heavy steel structures, but the cost can be prohibitive. An alternative is storage underground in caverns leached from salt deposits (fig. 11–15), in a manner similar to that used for crude oil storage.

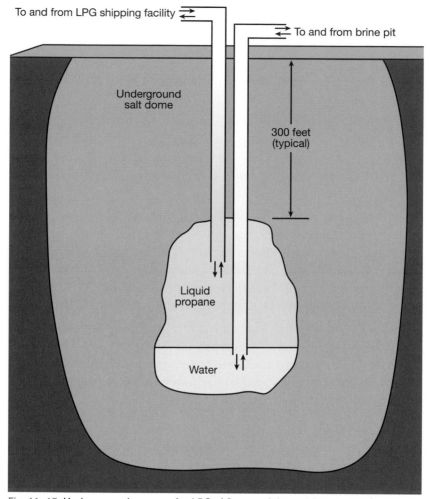

To and from LPG shipping facility

To and from brine pit

Underground
salt dome

300 feet
(typical)

Liquid
propane

Water

Fig. 11–15. Underground storage for LPGs (*Source:* Adapted from Leffler)

In the most common cavern-storage approach, a large quantity of the brine created during the leaching process is left in place at the bottom of the cavern. The petrochemicals or LPG are piped into the cavern, displacing the brine, which flows out through a second pipe and is stored in ponds at ground level. The liquid product floats on the brine that remains in the cavern.

To remove the product, brine is pumped from surface storage into the bottom of the cavern, pushing the product out through a separate pipe. At all times, however, the cavern is full of liquid, which prevents the walls from collapsing inward.

Standards for Refined Products

Standardization is essential in the creation of products from crude oil. For example, specifications for gasoline enable the design of engines for use in vehicles and aircraft in different nations around the world. Standardization also facilitates the movement of supplies from one region to another to meet shortages, encourages competition between refineries and distributors, and lets regulator set common rules for environmental controls for various fuel types.

Key parameters for crude and refined products include the following:
- Crude oil: API gravity, sulfur content, total acid number, distillation temperature profile (DTP)
- Gasoline: octane rating, Reid vapor pressure (RVP), oxygenate/sulfur/benzene content, DTP
- Jet fuel: flash point, smoke point, freezing point, DTP
- Diesel: cetane index, cloud point, sulfur content, DTP
- Heating oil: cloud point, sulfur content, DTP
- Residual fuel oil: viscosity, sulfur content, flash point, DTP

Many industry and governmental organizations participate in the development and enforcement of standards for petroleum products. These include:
- International Organization for Standardization (ISO)
- API
- ASTM International (formerly the American Society for Testing and Materials)
- SAE International (formerly the Society of Automotive Engineers)
- Environmental Protection Agency
- European Committee for Standardization (Comité Européen de Normalisation)
- Japan Standards Association
- Canadian General Standards Board

Military organizations also participate, such as the North Atlantic Treaty Organization, the U.K. Ministry of Defense, the U.S. Navy, and the U.S. Air Force.

Petrochemicals Production

Petrochemicals, derived from hydrocarbon molecules, account for about 6% of total global crude oil use. They are not used directly as fuel, but instead are used to enhance certain properties of fuels or as components in the manufacture of plastics, synthetic fibers, fertilizers, and hundreds of other niche industrial and consumer products.

The primary feedstocks used to produce petrochemicals in the United States and the Middle East are two NGLs: ethane and propane. In Asia, use of heavy naphtha and light fuel oil and gas oils (diesel and heating oil) is more common.

Petrochemical feedstocks come from an oil refinery or a gas separation plant. Butanes and light naphtha can come from either kind of facility. In addition, an oil refinery can provide heavy naphtha, light fuel oils, and gas oils.

Petrochemical plants are typically built close to refineries to be near feedstock sources. In addition, by-product streams from petrochemical production often can be best utilized by returning them to a refinery for processing. These factors have led to the development of several very large integrated refining/petrochemical complexes around the world. The largest, the Reliance Refinery, is operated by Reliance Petroleum, in Jamnagar, Gujarat, India, and has the capacity to refine 1.24 million b/d of crude oil.

Steam cracking is the principal method used to produce a major class of petrochemicals called *olefins*. As described above, the gas or liquid feedstock is mixed with steam, is heated to a very high temperature for a brief period, and then is cooled quickly to halt any further chemical reactions. The process breaks the molecules of the feedstock, and the hydrogen in the steam bonds with the cracked molecules to form new hydrocarbon molecules.

The main output from steam cracking are the olefins ethylene and propylene, which (along with benzene) are the principal petrochemical building blocks used to make products such as consumer plastics. Ethylene can be combined (usually by using a catalyst) to create polymers such as polyethylene, polyvinyl chloride (PVC), polystyrene, and polyester. It is also used to produce ethylene glycol (for making antifreeze and fibers). Because of its versatility, global demand for ethylene is high—in the range of 100 million tons per year. Most new plants can produce 1 million tons per year.

Propylene—the second most commonly produced petrochemical—is a by-product in the steam cracking of ethylene. It can also be produced in other ways: as a by-product of gas oil in FCC, through removal of hydrogen from propane, and through ethylene/butylene metathesis (a process in which molecules of the two substances switch places). Propylene is used to make polypropylene, acrylic acid, and acrylonitrile (for plastics manufacturing). The alkylation process also can convert propylene to a gasoline blendstock.

Important by-products of steam cracking for ethylene/propylene production are butadienes and butylenes. The former are used to make plastics resin and synthetic rubber; the latter are used as a gasoline blendstock and to produce the gasoline octane enhancer MTBE.

A final significant category of petrochemicals, the aromatics (collectively, BTEX), are all by-products of other processes, such as producing ethylene, reforming naphtha to produce gasoline blendstock, and producing raffinate (another gasoline blendstock) from the gas oil output from vacuum distillation. Benzene is used to make styrene and nylon; toluene is used to make solvents and urethane; and xylenes are used to make plasticizers, solvents, and gasoline octane enhancers.

12 Converting Natural Gas into Products

Natural gas is used in many ways in many countries around the world, as both a fuel and as a feedstock for making a range of other products.

Direct Gas Use as a Fuel

Natural gas is used primarily in combustion systems of various kinds to provide a range of energy services. These include generating electric power (in boilers and gas turbines), heating and cooling buildings, and running commercial and industrial equipment (e.g., water heaters, dryers, ovens, and furnaces).

Gas is also used directly as a vehicle fuel, in the form of *compressed natural gas* (CNG). Every major U.S. manufacturer of heavy-duty engines offers natural gas options for use in long-haul trucks, buses, and large vehicles used for mining, waste collection, and construction. Medium-duty vehicles—such as delivery trucks and airport shuttles—can use CNG instead of gasoline or diesel, as can light-duty vehicles (cars, light trucks, and taxis), chiefly in fleet operations. LNG has also been proved feasible for use in locomotives, ships, and tugboats.

Figure 3–9, in chapter 3, shows how much natural gas has been used in recent years in various sectors of the U.S. economy. In 2009, natural gas provided about 24% of total U.S. energy needs, and slightly more than half of all U.S. homes that year used natural gas as their main heating fuel.

Gas as a Petrochemical Feedstock

As noted in chapter 7 (see the "Gas Processing Operations" section), three commercially important chemicals are extracted from natural gas before the gas is sent to customers. These are ethane, propane, and butane.

Ethane

Ethane is used as a raw material in the petrochemical industry. It is processed to make ethylene and polyethylene—two building blocks for manufacturing many other products.

Ethylene is used in the manufacture of vinyl acetate (for use in paints and adhesives), polystyrene (used to make resins for rubber), and ethanol; it is also used to make ethylene oxide, which is itself a raw material for the production of ethylene glycol (antifreeze), as well as polyester fibers, film, and latex paints. Ethylene also can be combined with chlorine to yield vinyl chloride, which is used to make PVC, now widely used as a piping material.

Polyethylene is a plastic used in a variety of products. These include housewares, insulation, packaging films, and toys.

Propane and butane

If demand is low and the cost of recovery is high, propane and butane are simply flared or burned as fuel within a refinery. However, they also can be recovered and sold separately or as LPG in pressurized containers for cooking, heating, and some vehicle applications (e.g., forklifts). Butane is a main fuel source for many developing countries and is also used as a fuel for cigarette lighters.

Propane and butane are also used to manufacture propylene and butylene, two important chemical building blocks. Propylene is used to make propylene oxide (used to sterilize medical equipment and food products and to manufacture detergents), as well as propylene glycol (used in skin care lotions, industrial antifreeze, and hydraulic and brake fluid). Butylene is used in manufacturing additives that improve the quality of gasoline.

Gas-to-Liquids Technology

A process first developed in the 1920s by German scientists Franz Fischer and Hans Tropsch laid the foundation for a modern technology that converts natural gas directly into liquid fuels. The new gas-to-liquids (GTL) technology is the proprietary Shell Middle Distillate Synthesis process, developed by Royal Dutch Shell. Proved earlier at commercial scale at a Shell facility in Malaysia, the process has been scaled up to build the Pearl GTL plant in Ras Laffan Industrial City, in Qatar, which began commercial operation in mid-2011. The huge complex, with two production trains, is a joint venture with Qatar Petroleum. It will convert 1.6 bcf of natural gas per day from the giant offshore North Field into a variety of fuels and lubricants. In mid-June 2011, the joint venture announced sale of the first commercial shipment of GTL gas oil from the plant.

After removal of water, condensates, and NGLs, the purified natural gas (methane) is routed to a high-temperature gasifier. In that unit, the methane and oxygen are converted into *syngas*, a mix of hydrogen and carbon monoxide. Flowing through a series of vessels, the syngas then reacts with catalysts to form long-chain hydrocarbons that are subsequently broken into a range of smaller molecules to yield five products:

- GTL naphtha, a high-paraffin petrochemical feedstock
- GTL kerosene, for blending into aviation fuel
- GTL normal paraffins, for use in detergent production
- GTL gas oil, a diesel-type automotive fuel
- GTL base oils, for use in lubricants

Shell has reported that the $19 billion plant, which took five years to build, incorporates 35 years of research into the conversion process. At full production (slated for mid-2012), the plant is expected to produce 120,000 b/d of condensate, LPG, and ethane, plus 140,000 b/d of the five GTL products listed above.

GTL is seen as a potential alternative to LNG technology as a way to monetize natural gas resources. At a global level, GTL is still an emerging industry with a limited number of commercial plants operating, but several others were in planning or engineering stages as of mid-2010 (table 12–1).

Table 12–1. Global GTL plants

Location	Companies	Capacity, b/cda	Status
South Africa	Sasol	160,000	Operational
South Africa	PetroSA	22,500	Operational
Qatar (Ras Laffan)	Sasol/Qatar Petroleum	34,000	Operational
Qatar (Ras Laffan)	Shell/Qatar Petroleum	140,000: GTL products 120,000: LPG, ethane, condensate	Operational
Qatar (Ras Laffan)	ExxonMobil/Qatar Petroleum	154,000	Planning
Malaysia (Bintulu)	Shell	14,700	Operational
Nigeria (Escravos)	Chevron Nigeria/ Nigerian National Petroleum	34,000	Engineering
Papua New Guinea (Port Moresby)	Syntroleum	50,000	Planning
Uzbekistan (Qarshi)	Uzbekistan GTL	26,100	Feasibility study

Source: Adapted from *Oil & Gas Journal*, August 2, 2010.
a Barrels per calendar day.

Other Products Derived from Natural Gas

Natural gas is a raw material in the Haber process for making ammonia, which is used in turn to make agricultural fertilizer. Carbon black—used to reinforce rubber, make ink and batteries, and tint paints—comes from natural gas. Sulfur impurities removed from natural gas are often used as raw materials for producing agricultural chemicals.

13 Petroleum Industry Structure

Chapters 3–12 focused on the technologies and methods used by the petroleum industry to find, develop, produce, and upgrade oil and natural gas; to refine or otherwise transform it into useful products; and to store and move these vital fuels and their derivatives through the value chain and to the energy consumer.

The final section of this book (chaps. 13–15) changes focus, to discuss in broad terms several topics related to the business aspects of industry operations.

This chapter examines the structure of the industry, focusing on the kinds of companies and organizations involved in producing and refining crude oil. (Chapter 8 presented similar information about natural gas pipeline companies.)

Chapter 14 outlines some of the basic market and trading mechanisms and practices used to bring oil and natural gas to consumers. Chapter 15 suggests some of the major challenges that the industry may face in coming decades.

Major Oil and Gas Producers

The obvious place to start in examining the structure of the petroleum industry is the upstream side of the business (fig. 13–1)— that is, the companies that search for, develop, and extract oil and gas. Understanding the historical context is helpful, in view of the significant changes that have occurred over the past four or five decades; the next sections present a brief review. (For a more complete history of the industry, see, e.g., Yergin, 1991.)

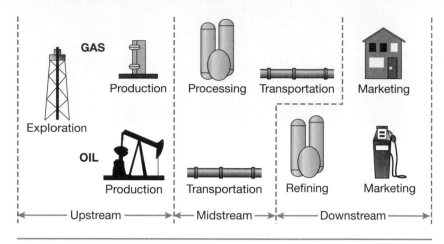

Fig. 13–1. General structure of the petroleum industry

The Seven Sisters, supermajors, and international oil companies

From the mid–1960s to the early 1970s, the global oil and gas trade was dominated by a handful of publicly owned companies based in the United States, the United Kingdom, and the Netherlands:

- Standard Oil of New Jersey (Esso)
- Standard Oil Company of New York (Socony)
- Standard Oil of California (Socal)
- Gulf Oil
- Texaco
- Royal Dutch Shell (Netherlands)
- Anglo-Persian Oil Company (UK)

This group was dubbed the *Seven Sisters* in the 1950s by Enrico Mattei, an Italian government administrator eager to have Italy's Eni (Ente Nazionale Idrocarburi, or National Fuel Trust) join the powerful club. After World War II, Mattei was instrumental in successfully transforming Agip (the state-owned oil company created by Mussolini) into Eni. Over the next half-century, mergers and acquisitions transformed the original Seven Sisters into the companies known today as ExxonMobil, Chevron, Royal Dutch Shell, and BP:

- Esso became Exxon, which renamed itself ExxonMobil when it acquired Mobil (formerly Socony) in 1999.

- Socal became Chevron, which acquired most of Gulf Oil in 1985 and then Texaco in 2001.

- Anglo-Persian Oil became Anglo-Iranian Oil in 1935 and then the British Petroleum Company in 1954. After acquiring Amoco (formerly Standard Oil of Indiana) in 1998 and making Atlantic Richfield Company (Arco) a subsidiary in 2000, British Petroleum officially changed its name to BP in 2001.

Operating in a manner not unlike that of a cartel, the Seven Sisters exerted considerable market power for many years over Third World oil producers. In 1973, the group controlled an estimated 85% of the world's petroleum reserves.

In recent decades, however, the dominance of the Seven Sisters and their successors—commonly referred to as *international oil companies* (IOCs)—has been challenged on several fronts. The cartel established in 1960 by OPEC has exerted increasing influence; the share of world oil production from OECD countries has declined; and state-owned oil companies have emerged in several developing countries.

In the late 1990s, the term *supermajors* became a more prevalent descriptor of the world's largest IOCs, a group that in most compilations includes Total (based in France) and ConocoPhillips as well as ExxonMobil, Chevron, Shell, and BP. Total acquired Belgium's Petrofina in 1999, merged with the French firm Elf Aquitaine in 2000 and then with Spain's Compania Espanola de Petroles (Cepsa) in 2001.

Conoco (formerly Continental Oil and Transportation) merged with Phillips Petroleum in 2002 to form ConocoPhillips. Burlington Resources joined ConocoPhillips in 2006. (In mid-2011, ConocoPhillips said it would spin off its refining assets by mid-2012, investing cash from the transaction to invest in oil exploration.)

The supermajors resulted from the mergers and acquisitions noted above, as well as others. Undertaken in response to a severe deflation in oil prices, those actions aimed to achieve larger economies of scale, hedge against oil-price volatility, and reinvest large cash reserves.

Interestingly, several of today's IOCs began life as government-sponsored firms but were subsequently privatized. These include Petro-Canada; the Oil and Natural Gas Corporation of India (ONGC); and Russia's Gazprom, Lukos, Surgutneftegaz, and Rosneft. Although they are still significant players on the world energy stage, as a group, the supermajors controlled only about 6% of global oil and gas reserves in 2009 (fig. 13–2).

TOP 20 COMPANIES BY REMAINING RESERVES

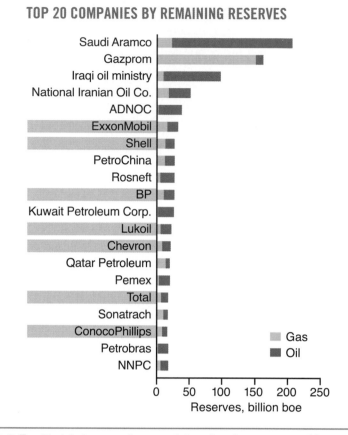

Fig. 13–2. Top 20 global companies: remaining oil and gas reserves (*Source: Oil & Gas Journal*, March 7, 2011 [2009 data from Wood Mackenzie])

National oil companies

Beginning in the 1960s, in response to the power of the IOCs and concern about control over energy resources and markets, several nations took action to create or sponsor new entities. At the forefront were the members of OPEC, who took as their examples the actions of Mexico, which created Petróleos Mexicanos (Pemex) in 1938, and of Iran, which created the National Iranian Oil Company in 1951. This initiative gained additional momentum when the Arab oil embargo shocked global petroleum markets in 1973.

In general terms, these new entities, simply called *national oil companies* (NOCs), were typically built from existing petroleum ministries or from the assets (purchased or nationalized) of companies operating in the country. Solely owned and operated by their respective governments, NOCs conducted their business with much less transparency than did the IOCs.

The following are examples of NOCs founded in this way, along with the date(s) of creation:

- Sonatrach (Algeria), 1963–71
- Libyan National Oil, 1968–70
- Saudi Arabian Oil (Saudi Aramco), 1973–80
- Abu Dhabi National Oil, 1971
- Petroliam Nasional Berhad (Petronas), Malaysia, 1974
- Petróleos de Venezuela (PdVSA), 1975
- Kuwait National Petroleum, 1975
- Nigerian National Petroleum, 1977

In 2011, approximately 88% of global oil and gas reserves were controlled by state-owned oil companies, primarily located in the Middle East.

Hybrid companies

A third kind of major petroleum company became more prominent starting about 2000. This hybrid entity, part public and part government owned, is sometimes referred to as a *government-sponsored enterprise* (GSE). Examples include the following:

- Petroleo Brasileiro SA (Petrobras) (Brazil)
- OMV Petrom (Romania)
- Statoil-Hydro (Norway), now Statoil

In the late 1990s, China began a major restructuring of its energy industries to transform the formerly state-owned China National Petroleum Corporation (CNPC)—which focused on domestic exploration and production—and the China Petroleum and Chemical Corporation (Sinopec)—which managed petroleum refining and distribution. The goal was to transfer some of the assets of the two corporations into vertically integrated firms that would be jointly owned by the government and investors. The entities that emerged from this reorganization are the China National Offshore Oil Corporation (CNOOC), Sinopec itself, and PetroChina, which is a spin-off of CNPC.

A report by IEA in early 2011 indicated that Chinese oil companies operate with a high degree of independence from the government. By late 2010, Chinese oil companies had *equity oil* (i.e., control over the disposition of their share of production) in 20 of the 31 countries where they operated.

Even though GSEs are structured like private-sector corporations, the factors shaping their behavior are complex. Investors (and governments) expect them to be profitable, but they also must serve national interests (e.g., securing oil imports and/or developing domestic energy resources). For example, Norway's Statoil had approval to leverage its expertise in deepwater development technology to pursue projects (alone or in partnerships) in other regions.

Independent Producers

Independent producers, also called *independents*, are smaller and not vertically integrated in the way that many IOCs, NOCs or GSEs are. Independents play a significant role in the upstream oil and gas business, operating chiefly in the United States but in several foreign countries as well.

Independents receive nearly all of their revenues from production at the wellhead. Chiefly in the exploration and production side of the petroleum business, they engage in little marketing or refining activity. For a U.S. firm to be considered an independent producer, Internal Revenue Service regulations stipulate that its refining capacity must average less than 75,000 b/d or its retail sales must be less than $5 million per year.

A variety of organizations represent the interests of independent producers. Operating at the national level is the Independent Petroleum Association of America (IPAA).

However, many state and regional groups provide similar support. One example is the Petroleum Technology Transfer Council (PTTC), which lists as members more than 40 state/regional producer organizations, including producers and royalty owners associations; oil and gas associations; landowners associations; petroleum councils; marginal well commissions; and alliances of energy producers. PTTC is a national not-for-profit organization founded in 1994 and managed by the American Association of Petroleum Geologists. PTTC seeks "to connect independents with the technology and knowledge to safely and responsibly develop the nation's oil and gas resources."[1]

Through interactions with PTTC, their regional associations, and various research organizations (e.g., Gas Research Institute, now GTI), independent producers have also helped to develop and evaluate important new technologies (e.g., horizontal drilling and hydraulic fracturing) for improved exploitation of petroleum resources.

IPAA commissioned a study by IHS Global Insight (USA) evaluating the role of independents in the U.S. economy.[2] The study looked only at the upstream sector, evaluating the activities of 18,000 independent producers operating in 32 states. It found that for the year 2010, these independents:

- Drilled 95.3% of onshore and 69.7% of offshore oil and gas wells
- Produced 86.3% of onshore U.S. natural gas and 69.4% of offshore production
- Operated most U.S. marginal wells (yielding less than 16 boe per day), which account for some 80% of the nation's 800,000 producing wells

IPAA reported that independents hold 80% of the shallow-water leases in the Gulf of Mexico, and 75% of deepwater leases. They highlighted that independents play a larger role both in the Lower 48 states and in near-offshore areas, as IOCs and large integrated producers pursue development of larger fields in frontier areas, such as ultra-deepwater.

Prominent publicly held U.S.-based independent oil and gas producers include the firms Anadarko Petroleum, Barrett Resources, Cabot Oil and Gas, Chesapeake Energy, Devon Energy, EOG Resources, Forest Oil, Harvest Natural Resources, Noble Energy, Plains Exploration and Production, Quicksilver Resources, Swift Energy, Ultra Petroleum, and Ward Petroleum Group.

Independents based in other countries are also exerting an impact on global oil and gas production. Major foreign-based independents include Bankers Petroleum, Petrobank Energy and Resources, Talisman Energy, and Vermillion Resources (Canada); British Gas and Cairn Energy (UK); Carnarvon Petroleum and Nido Petroleum (Australia); Maersk Oil (Denmark); and Japex (Japan). To that list must now be added Melbourne, Australia–based BHP Billiton, the world's largest mining company, which in July 2011 agreed to buy U.S. independent Petrohawk Energy for $12.1 billion; Petrohawk was a major player in U.S. shale gas development.

Service Companies

Oil and gas producers large and small typically turn to specialty firms, called *service companies*, to conduct the hands-on work of building and operating drilling rigs. Service companies provide lubricants, other fluids/ materials, and pumps and other equipment, as well as the expertise and tools needed to assess formation characteristics and well performance. As the price of oil goes up (as it did in 2010–11), demand for oil field services rises as well.

The *Financial Times* prepares an annual list of the world's biggest companies, called the *FT Global 500*. In the oil equipment and services sector, eight companies made the list for 2011 (table 13–1). A major service company that did not make the list is Weatherford International, based in Switzerland.

National Oilwell Varco made the biggest jump in the listing for 2011 (table 13–1), based in part on its provision of equipment needed for hydraulic fracturing of shale to extract oil and gas. Such technology found extensive use in the United States and elsewhere in 2010 and 2011. In fact, oil field service companies were instrumental in pioneering the technique of horizontal drilling to extract gas and oil from shale.

Service companies apply specialized skills and technology to help halt declining output from producing fields, maintain aging wells and related equipment, or search for hydrocarbons in challenging environments.

Table 13–1. FT Global 500, oil equipment and services, 2010–11

Company	Global rank 2011	Global rank 2010	2011 market value,[a] billion USD
Schlumberger (US)	41	71	126.8
Halliburton (US)	176	279	45.5
National Oilwell Varco (US)	262	478	33.4
Baker Hughes (US)	274	—	31.9
TransCanada (Canada)	321	312	28.3
Transocean (US)	363	273	25.2
Enbridge (Canada)	390	438	23.6
Saipem (Italy)	391	474	23.5

Source: Adapted from the *Financial Times*, 2011 FT Global 500 list.
[a] Market values in U.S. dollars as of March 31, 2011, based on data from Thomson ONE Banker, Thomson Reuters Datastream, and individual companies.

IOC-NOC Interactions

As IOCs and NOCs (as well as hybrid GSEs) pursue their various business and political interests, new kinds of relationships are developing. This is driven by the recognition that NOCs now are estimated to control more than 85% of global oil and gas reserves. Some NOCs rival the majors in size and scope and even compete directly with IOCs for access to reserves. Complicating the situation, other players—including major oil field services firms, private equity firms, sovereign wealth funds, and third-party energy marketers—have taken on roles previously dominated by IOCs.

A 2009 analysis by Ernst and Young postulated that this new reality has shifted the power structure of the global petroleum industry, so that IOCs must now convince NOCs and host governments that they can provide value beyond just finding and producing oil and gas.[3] Particularly in the planning and execution of expensive, complex, high-risk megaprojects, IOCs offer vertically integrated skills, knowledge, and technology that can help resource-rich countries to capitalize on their hydrocarbon reserves. Industry experts have predicted that global capital spending on new projects could reach $400 billion during 2008–15, with much of it slated for megaprojects typically costing $5 billion each. The challenge, according to Ernst and Young, will be to customize their offerings to address the unique needs, beyond exploration and production, of NOCs. This may include training a local workforce, developing related processing and delivery infrastructure, and assisting with other economic development efforts.

Energy industry news from late 2009 to mid-2011 was replete with examples of the ways in which IOCs, NOCs, and other players are interacting. The following are a select few:

- Major Abu Dhabi sovereign wealth fund (Ipic) announced a plan to take full ownership of Cepsa, Spain's second-largest oil company, buying the stake previously held by France's Total.
- BP negotiated (unsuccessfully) with Russia during early 2011 on a $16 billion deal that would have given BP access to Russia's Arctic oil fields and would have authorized BP and state-owned Rosneft to swap shares.
- Shell and Russia's Gazprom formed an alliance to strengthen cooperation on projects in Russia and work together in other nations.

- Shell signed a $12.5 billion agreement with Iraq to capture flared gas from that country's oil fields. Mitsubishi announced that they would also participate in the project.

- Korea National Oil signed a $1.5 billion deal with U.S. firm Anadarko for one-third of the latter's interest in part of the Eagle Ford shale formation (Texas).

- Thailand's biggest oil company, PTT Exploration and Production, paid $2.3 billion to Statoil for a 40% stake in a Canadian oil-sands project.

- Companies expecting to start drilling in 2011 off Cuba included Venezuela's PdVSA (using a drilling rig built in China), Spain's Repsol YPF, Malaysia's Petronas, Russia's Gazprom, and India's ONGC.

- Occidental Petroleum won the lead contract to develop the Shah sour gas project in the United Arab Emirates, sharing a stake with Abu Dhabi National Oil.

- BP signed a 30-year production-sharing deal with Socar (Azerbaijan's NOC) to develop the Shafaq-Asiman offshore gas field, southeast of Baku.

- TNK-BP, the Russian venture half-owned by BP, agreed to pay $1.8 billion to buy BP's pipeline and production assets in Venezuela and Vietnam.

China's global outreach has been particularly striking in recent years. An assessment for the *Financial Times* showed that from January through early November 2010, that country's national oil companies spent more than $24 billion on overseas acquisitions—accounting for 20% of all global deals in the upstream oil and gas sector for the period.[4] An analysis by Wood Mackenzie concluded that China's three largest NOCs would realize a record one million boe/day in net overseas oil and gas production in 2010.[5]

China's initiatives from late 2010 to mid-2011 included the following:

- PetroChina agreed to pay $5.4 billion for a 50% stake in the Cutback Ridge shale gas project of Encana in western Canada. (It is China's largest investment in shale gas to date).

- PetroChina and Sinopec spent $1.9 billion and $4.65 billion, respectively, for stakes in Canadian oil sands projects.

- PetroChina acquired a 50% stake in major private UK chemicals production firm Ineos to refine and trade oil at Scottish and French refineries.

- CNPC won a bid to expand an oil refinery in Cuba (a $6 billion deal) that is jointly owned by PdVSA (Venezuela) and Cubapetroleo.

- CNPC agreed to buy Occidental Petroleum's operations in Argentina for $2.45 billion.

- CNOOC (China's largest offshore oil and gas producer) signed two deals with Chesapeake Energy, each worth just over $1 billion, regarding U.S. acreage holding shale oil and gas resources.

- CNOOC and Bridas Energy Holdings (a private Argentine company) agreed to pay $7 billion for BP's 60% stake in Argentina's Pan American Energy. (CNOOC previously bought a 50% stake in Bridas for $3.1 billion.)

- Sinochem submitted the winning $3.1 billion bid for a share of the Peregrino oil field off the coast of Brazil.

- Statoil worked to finalize a deal to explore for shale gas in China as Sinopec and PetroChina also sought help from BP, Chevron, and Shell to develop domestic shale gas.

OPEC

OPEC is a major force in the global trade of crude oil. The organization marked its 50th anniversary in 2011. Typically OPEC meets twice each year to assess current and projected world oil supply/demand and prices, as well as to coordinate the actions of its members in regard to production and pricing.

OPEC's mission is to "coordinate and unify the petroleum policies of its Member Countries and ensure the stabilization of oil markets in order to secure an efficient, economic and regular supply of petroleum to consumers, a steady income to producers and a fair return on capital for those investing in the petroleum industry."[6]

OPEC has historically used both pricing and production controls to achieve its goals. This entailed establishing an official posted price for

OPEC oil, increasing or reducing overall production by its members, and (starting in 1983) setting production quotas for individual member countries.

In recent years, OPEC member nations have produced about 40% of the world's crude oil and 18% of its natural gas. In addition, OPEC's oil exports represent about 60% of all the oil traded internationally.

OPEC history

The genesis of OPEC can be traced to a meeting in 1949 of officials from Venezuela, Iran, Iraq, Kuwait, and Saudi Arabia. These major oil producers sought to exchange views as the global energy market began to reshape itself following World War II. By establishing better communications among themselves, the group of oil-producing nations were able to form a more united front in their interactions with the IOCs, who were licensed to produce and export oil from their countries. (Recall the discussion about the Seven Sisters.) In particular, the group was rankled by the decisions of IOCs that had set relatively low prices for crude oil from Venezuela and the Middle East.

Price reductions made in 1959 and 1960 prompted a mid–September 1960 conference in Baghdad, attended by the same five nations. It was at this meeting, the First OPEC Conference, that it was decided to establish OPEC as a permanent intergovernmental organization. At the Second OPEC Conference (Caracas, 1960), the group decided to base its secretariat in Geneva, Switzerland. (In 1965, it was moved to Vienna, Austria.) In 1973, the aggregate output of OPEC member nations accounted for more than 50% of worldwide crude oil production.

OPEC membership

As of mid–2011, OPEC membership had grown to 12 states (table 13–2). In addition to the five founder members, there are seven full members that have joined since 1961. (Two former full members, Gabon and Indonesia, ended their memberships in 1995 and 2008, respectively.)

According to OPEC, any country with "a substantial net export of crude petroleum, which has fundamentally similar interests to those of Member Countries, may become a Full Member of the Organization, if accepted by a majority of three-fourths of Full Members, including the concurring votes of all Founder Members."[7] OPEC also has associate member countries that do not qualify for full membership but are admitted under such special conditions.

Table 13-2. OPEC membership, June 2011

Nation[a]	Year joined
Iran*	1960
Iraq*	1960
Kuwait*	1960
Saudi Arabia*	1960
Venezuela*	1960
Qatar	1961
Libya	1962
United Arab Emirates	1967
Algeria	1969
Nigeria	1971
Ecuador[b]	1973
Angola	2007

Source: OPEC.
[a] Founder members are identified with asterisks. Others are full members.
[b] Ecuador suspended its membership in 1992 and reactivated it in 2007.

OPEC outlook

Analysts debate how successful OPEC has been and how effective it will be in an increasingly chaotic global economy. Compliance by its members with production quotas has been sporadic at best, because of the lack of an effective enforcement mechanism. Actual total OPEC oil production has typically exceeded the quota-defined ceiling by 4% since 1983 (and occasionally by more than 15%).

The following factors also impede effective operation:

- Lack of accurate and timely information about future changes in oil demand and the availability of non-OPEC oil supplies (despite regular forecasts from such agencies as the International Energy Agency and OPEC itself)

- Time delay (typically several years) in observing the full impact of a price change

- Nonalignment of the interests of members—owing to differences in population, magnitude of oil reserves, and the role of oil revenues in addressing domestic social and development needs

Some members would prefer to maximize their revenue from oil taxes and sales, supporting higher oil prices. Others fear that high prices could (*a*) further damage an already weak global economy and destroy demand or (*b*) spur innovations in alternative-energy or conservation technologies. Members with relatively modest reserves express concern that high production rates would deplete their petroleum legacy too rapidly.

Also affecting the success of OPEC are geopolitical disruptions and other unanticipated events. Examples include:

- Military conflict—in particular, the 1973 Arab-Israeli War (1973) and the Persian Gulf Wars (begun in 1990 and 2003)

- Surge in Asian oil demand, which tightened oil markets in 2005

- Political protests—from Tunisia and Egypt to Libya and Syria— in the "Arab Spring" uprisings of 2011 (see below)

In recent years, signs of internal OPEC discord have begun to emerge, reflecting the changing political priorities and economic fortunes of several prominent members. A key issue for OPEC is the timing and scope of Iraq's planned return to significant oil production levels. By some estimates, with the help from foreign firms, Iraq could raise its output to 12–14 million b/d, compared to just 2.5 million in 2010.

Such a major increase could lead OPEC leaders to hold down overall production (aiming to prevent an oil price drop), reducing production quotas for other OPEC members, to "make room for" Iraqi output. (Iraq has not had an OPEC production quota since Saddam Hussein's regime invaded Kuwait in 1990.)

At the June 2011 OPEC meeting in Vienna, a "lack of harmony" was on display, as reported in *The Economist*.[8] Considering a proposal to increase production by 1.5 million b/d, members split evenly, so quotas remained at the levels set in December 2008. The price of Brent crude spiked on the news, hitting $118.43 per barrel.

Those arguing for higher production pointed to the loss of about 1.2 million b/d from Libya, where civil unrest continued after flaring in February. Others asserted that a continued high oil price was needed to generate revenues needed to help governments shelter citizens against sharply rising costs of food and other commodities.

On the one hand, the cartel noted that the devastating March 2011 earthquake and resultant tsunami in Japan drove down oil demand midyear, as did continuing economic uncertainty in the United States and elsewhere. On the other hand, the continued growth of China's economy partially offset downturns in oil demand elsewhere.

Further confounding the June 2011 discussions were the following:

- Information suggesting possible softening of global oil demand through 2012, which would negate the need for more supply

- Ill will due to Qatari support of rebel fighters in Libya and Saudi support of Bahrain's suppression of a Shiite uprising (which angered Shiite Iran)

Energy markets and political dynamics will, of course, continue to evolve in coming years. Some observers of recent events suggest that OPEC's power may be waning and that non-OPEC nations (e.g., Russia and Kazakhstan) could be poised to take a more prominent role on the world energy stage. At the same time, the following must be acknowledged:

- OPEC members produce a significant share of daily global oil output.

- Several OPEC members (particularly Saudi Arabia) have sufficient spare capacity to vary production and thereby shape the market.

In fact, just days before the June 2011 OPEC meeting, Saudi Aramco said it would speed development of the supergiant Manifa oil field, aiming to produce at its maximum rate (900,000 b/d) by 2014—a decade earlier than previously planned. Furthermore, a week after the meeting, the government said it would boost total national production to more than 10 million b/d by July—the highest level in 25 years and close to the kingdom's total production capacity of 12.5 million b/d.

In a mid-2011 analysis, EIA projected that the OPEC share of total global crude and liquid fuels production would increase slightly in the near term, from 40.5% in 2010 to 40.9% in 2011 (table 13–3). Economic, technical, and political factors still to emerge will certainly shape OPEC's role in the years beyond that time frame.

Table 13–3. Global crude oil and liquid fuels supply, 2010–2012

Group	Production, million b/d			% of total world		
	2010	2011	2012	2010	2011	2012
Non-OPEC nations	51.70	52.29	52.78	59.5	59.6	59.1
OPEC nations:						
Crude oil only	29.77	29.40	30.06	34.3	33.5	33.7
Other liquids	5.39	6.00	6.44	6.2	6.8	7.2
OPEC total	35.16	35.40	36.50	40.5	40.4	40.9
World total	86.86	87.69	89.28			

Source: EIA, *Short Term Energy Outlook*, June 7, 2011; adapted from table 3a.

Refiners

EIA reported that there were 142 operable refineries in the United States as of June 2011 (tables 13–4 and 13–5). Of those, 114 had a capacity of at least 25,000 barrels per calendar day (b/cd); all of the top 65 have capacities of at least 100,000 b/cd. Many of the largest are owned by well-known IOCs—the largest, in Baytown, Texas, by ExxonMobil.

Total U.S. refining capacity (counting only atmospheric crude oil distillation units) has been relatively stable in recent years, at about 16 million b/cd. Previously, capacity grew consistently for about 10 years starting in the early 1970s, rising to 19.4 million b/cd in 1981. At that point, however, dropping oil demand put pressure on refineries, and the first to shut down were those that had only simple distillation systems

Table 13–4. Largest U.S. oil refineries ranked by capacity, June 2011

Company[a]	Location	Capacity,[b] b/cd
ExxonMobil Refining and Supply	Baytown, TX	560,640
ExxonMobil Refining and Supply	Baton Rouge, LA	502,000
Hovensa (Hess/PdVSA joint venture)	Kingshill, Virgin Islands	500,000
Marathon Petroleum	Garyville, LA	464,000
Citgo Petroleum	Lake Charles, LA	427,800
BP Products North America	Texas City, TX	406,570
BP Products North America	Whiting, IN	405,000
WRB Refining	Wood River, IL	362,000
ExxonMobil Refining and Supply	Beaumont, TX	344,500
Sunoco (R&M)	Philadelphia, PA	335,000
Chevron USA	Pascagoula, MS	330,000
Deer Park Refining	Deer Park, TX	327,000
Premcor Refining Group	Port Arthur, TX	292,000
Flint Hills Resources	Corpus Christi, TX	290,078
Motiva Enterprises	Port Arthur, TX	285,000
Houston Refining	Houston, TX	280,390
Chevron USA	El Segundo, CA	273,000
Flint Hills Resources	St. Paul, MN	262,000
BP West Coast Products	Los Angeles, CA	253,000
ConocoPhillips	Belle Chasse, LA	247,000
ConocoPhillips	Sweeny, TX	247,000
Chevron USA	Richmond, CA	245,271
ConocoPhillips	Westlake, LA	239,400
ExxonMobil Refining & Supply	Joliet, IL	238,600
ConocoPhillips	Linden, NJ	238,000

Source: EIA, Ranking of U.S. Refineries, June 2011.
[a] Includes only refineries with atmospheric crude oil distillation capacity.
[b] Based on capacity data from refineries as of January 1, 2011.

and little capability for subsequent upgrading (i.e., improving their downstream processing capability).

Additional refineries closed in the late 1980s and during the 1990s. At the same time, however, refiners found ways to improve the efficiency of the crude oil distillation units that remained in service. This process, called *debottlenecking*, better matches flow capacity among different units and makes wider use of computer control. Motivated by environmental mandates (as well as economics), refiners also improved their upgrading capability. As a result, the capacity of downstream units was no longer the constraining factor on the amount of crude oil processed (or run) through the crude oil distillation system.

Crude oil runs climbed through the 1990s, and an associated parameter, *capacity utilization*, exceeded 90% from the mid–1990s through

Table 13–5. Smallest U.S. oil refineries ranked by capacity, June 2011

Company[a]	Location	Capacity,[b] b/cd
Ergon West Virginia	Newell, WV	20,000
Petro Star	North Pole, AK	19,700
Western Refining Southwest	Bloomfield, NM	16,800
ConocoPhillips Alaska	Prudhoe Bay, AK	15,000
San Joaquin Refining	Bakersfield, CA	15,000
Age Refining	San Antonio, TX	14,021
Wyoming Refining	New Castle, WY	14,000
Calumet Lubricants	Cotton Valley, LA	13,020
BP Exploration Alaska	Prudhoe Bay, AK	12,780
Ventura Refining & Transmission	Thomas, OK	12,000
Hunt Southland Refining	Sandersville, MS	11,000
Silver Eagle Refining	Woods Cross, UT	10,250
American Refining Group	Bradford, PA	10,000
Montana Refining	Great Falls, MT	10,000
Greka Energy	Santa Maria, CA	9,500
Lunday Thagard	South Gate, CA	8,500
Calumet Lubricants	Princeton, LA	8,300
Martin Midstream Partners	Smackover, AR	7,500
Valero Refining	Wilmington, CA	6,300
Somerset Energy Refining	Somerset, KY	5,500
Goodway Refining	Atmore, AL	4,100
Garco Energy	Douglas, WY	3,600
Silver Eagle Refining	Evanston, WY	3,000
Tenby	Oxnard, CA	2,800
Foreland Refining	Ely, NV	2,000

Source: EIA, *Ranking of U.S. Refineries*, June 2011.
[a] Includes only refineries with atmospheric crude oil distillation capacity.
[b] Based on capacity data from refineries as of January 1, 2011.

2004. (Capacity utilization is the total amount of crude oil, unfinished oils, and NGLs run through a crude oil distillation unit divided by the capacity of the unit. A unit operating all-out has a capacity utilization of 100%.) EIA reported that utilization was more than 95% of capacity in 1997 and 1998 and averaged above 90% through 2005.

The U.S. recession caused lower petroleum demand and pushed utilization capacity down to 83% in 2009. That figure recovered slightly, to approximately 86% in 2010.

Some in the industry have talked about 1995–2005 as a golden age for refiners. Notable improvements in performance were achieved despite a drop in the number of U.S. refineries—from just over 300 in 1981, to 192 in 1985, and to 146 in 2003.

In June 2011, EIA reported that U.S. oil-refining capacity increased to 17.7 million b/cd during 2010—its highest level in nearly three decades, with 148 refineries operating in January 2011. The capacity increase over 2009 was 152,000 b/cd.

The Gulf Coast has more than twice the crude distillation capacity of any other region in the United States. The difference is even greater for downstream processing capacity. The Gulf Coast region is also the national leader in supply of refined products. For example, more than half of East Coast needs—and 20% of Midwest needs—for light products such as gasoline, heating oil, diesel, and jet fuel come from the Gulf Coast.

Independent refiners

As a general rule, independent refiners have little involvement in oil production, choosing instead to purchase crude oil produced by others and process it into finished products. They may also own wholesale and retail marketing outlets or sell their products to marketing companies.

During the 1990s, the role of independent refiners grew substantially, largely as the result of refinery purchases from IOCs seeking to streamline and realign their positions. In more recent years, independent refiners themselves underwent a period of consolidation; mergers and acquisitions scrambled the refinery ownership roster but had little impact on overall refined product supply.

Many independent refiners are members of the National Petrochemical and Refiners Association. Valero Energy was the largest independent refiner in the United States in early 2011; second was Marathon Petroleum. Other prominent independents include Frontier Oil, Holly, and Sunoco (R&M).

World refining capacity

Broadly speaking, refining developed in consuming areas because it was cheaper to move crude oil than to move petroleum products. In addition, proximity to consuming markets made it easier to respond to weather-induced spikes or seasonal shifts in demand. Thus, while the Middle East is the largest oil-producing region, the bulk of oil refining takes place in the United States, Europe, and Asia.

One historical exception was establishment of a significant refining center in the Caribbean to supply heavy fuel oil to U.S. East Coast power, heat, and electric generation markets. However, as the demand for this heavy (residual) fuel oil slipped, the economic rationale for those dedicated refineries could no longer be upheld. Most refineries meet local or regional demand first, using exports as a temporary method to balance supply and demand. Exceptions to this rule are the Caribbean refineries described above, as well as refineries in the Middle East and Singapore.

At the start of 2012, the largest concentration of refining capacity (including crude distillation as well as subsequent upgrading) was in North America, primarily in the United States (table 13–6). Note that Asia held the lead with regard to crude distillation only. Because North America (and the United States in particular) uses a large majority of its oil in the form of gasoline, the region has by far the largest downstream capacity (necessary to maximize gasoline output).

Table 13–6. Regional view of global refining capabilities, January 1, 2012

Region:	Africa	Asia	Eastern Europe	Middle East	North America	South America	Western Europe	Total
No. of refineries	45	164	89	44	148	66	99	655
Crude distillation	3,218	24,918	10,369	7,277	21,246	6,595	14,431	88,055
Vacuum distillation	510	4,833	3,903	1,982	9,309	2,847	5,678	29,062
Catalytic cracking	210	3,208	877	358	6,513	1,311	2,216	14,693
Catalytic reforming	458	2,209	1,474	660	4,126	402	2,138	11,468
Catalytic hydrocracking	62	1,242	330	597	1,936	132	1,189	5,489
Catalytic hydrotreating	833	10,223	4,274	2,047	16,371	1,904	10,077	45,730
Coke production	1.84	20.45	12.57	3.30	133.71	24.64	12.61	209.12

Source: Adapted from *Oil & Gas Journal*, December 5, 2011, p. 36.
Note: All capacity figures are given in thousand b/cd, except coke, which is given in thousands of tonnes per day.

An assessment of trends in worldwide refining by *Oil & Gas Journal* in late 2010 found that total global crude distillation capacity increased for that year to more than 88.2 million b/cd.[9] Other key points from the assessment follow:

- A ninth consecutive year of increase in crude refining capacity occurred in 2010—a jump of 1 million b/cd over 2009 (fig. 13–3). The 2010 growth was smaller than for 2009 but well ahead of growth for 2008, 2007, and 2006.

- Capacity growth for 2010 was almost entirely in Asia and the Middle East. Asian capacity grew by nearly 1.5 million b/cd (largest of any region), with China and India leading the way.

- The number of refineries grew in 2010, but at a slower pace than for 2009. The growth in 2009 ended a decade of contraction in refinery count.

- Three new refineries started up in 2010, all in the Asia–Pacific region, as had also been the case in 2009.

- North America's capacity remained virtually flat.

- Western Europe's capacity contracted as companies closed inefficient plants or took other actions to meet changing market demands.

- Other regions experienced little or no net gain or loss in stated capacity, although activity in the Middle East was brisk.

WORLDWIDE REFINING*

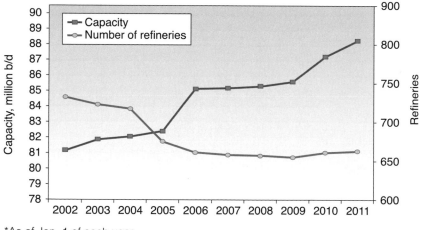

*As of Jan. 1 of each year.

Fig. 13–3. Trends in operating refineries and global refining capacity, 2001–2011 (*Source: Oil & Gas Journal*, December 6, 2010)

Largest refiners and refineries

Table 13–7 lists the 25 refining companies that own the most worldwide capacity. Table 13–8, 13–9, and 13–10 list those refiners whose plants have more than 200,000 b/cd of capacity in Asia, the United States, and Western Europe, respectively. Table 13–11 presents information about the largest refineries in the world.

Table 13–7. Ranked list of the world's largest refiners

Rank as of Jan. 1 2012	2011	Company	Crude capacity,[a] thousand b/cd
1	1	ExxonMobil	5,788
2	2	Royal Dutch Shell	4,194
3	3	Sinopec	3,971
4	4	BP	3,322
5	8	Valero Energy	2,776
6	7	PdVSA	2,678
7	9	China National Petroleum	2,675
8	5	ConocoPhillips	2,568
9	6	Chevron[b]	2,560
10	11	Saudi Aramco	2,451
11	10	Total	2,314
12	12	Petrobras	1,997
13	13	Pemex	1,703
14	14	National Iranian Oil	1,451
15	15	JX Nippon Oil & Energy	1,423
16	16	Rosneft	1,293
17	17	OAO Lukoil	1,217
18	18	Marathon Petroleum	1,193
19	24	SK Innovation	1,115
20	19	Repsol YPF	1,105
21	20	Kuwait National Petroleum	1,085
22	21	Pertamina	993
23	22	Agip Petroli	904
24	25	Flint Hills Resources	816
25	23	Sunoco (R&M)	675

Source: Oil & Gas Journal, Dec. 5, 2011, p.32.
[a] Figures include partial interests in refineries not wholly owned by the company.
[b] Includes holdings in Caltex Oil and GS Caltex.

Table 13–8. Companies in Asia with at least 200,000 b/cd refining capacity, January 1, 2012

Company	No. of refineries	Crude capacity,[a] thousand b/cd
1. Sinopec	27	3,971
2. China National Petroleum	25	2,675
3. ExxonMobil	10	1,938
4. JX Nippon Oil & Energy	7	1,423
5. Royal Dutch Shell	13	1,341
6. Indian Oil	11	1,315
7. Reliance Petroleum	2	1,240
8. SK Innovation	1	1,115
9. Pertamina	8	1,012
10. Chinese Petroleum	3	770
11. GS Caltex	1	750[b]
12. Tonen/General Sekiyu Seisei KK	4	628[c]
13. Idemitsu Kosan	4	608
14. Chevron	6	583
15. Cosmo Oil	4	565
16. S-Oil	1	565[d]
17. Formosa Petrochemical	1	540
18. BP	4	352
19. Saudi Aramco	6	328
20. Hyundai Oil Refinery	3	319
21. Hindustan Petroleum	2	298

Source: Adapted from *Oil & Gas Journal*, December 5, 2011, p. 34.

Note: Asia includes Australia, Bangladesh, Brunei, China (and Taiwan), India, Indonesia, Japan, Malaysia, Myanmar, New Zealand, North Korea, Pakistan, Papua New Guinea, the Philippines, Singapore, South Korea, Sri Lanka, and Thailand.

[a] Figures include partial interests in refineries not wholly owned by the company.

[b] Includes Caltex's 50% stake.

[c] Includes ExxonMobil's 50% stake.

d Includes Saudi Aramco's 35% stake.

Table 13–9. Companies in the United States with at least 200,000 b/cd refining capacity, Jan. 1, 2012

Company	No. of refineries	Crude capacity,[a] thousand b/cd
1. ConocoPhillips	13	2,276
2. Valero Energy	12	2,096
3. ExxonMobil	7	2,042
4. BP	6	1,486
5. Marathon Oil	6	1,193
6. Royal Dutch Shell	8	1,011[b]
7. Chevron	5	955
8. PdVSA	4	849[c]
9. Flint Hills Resources (Koch Industries)	3	816
10. Motiva Enterprises[d]	3	772
11. Sunoco (R&M)	3	675
12. Tesoro	7	658
13. Saudi Aramco	3	410[e]
14. Encana	2	276
15. LyondellBasell	1	268
16. Alon USA	3	241
17. Husky Energy	2	238

Source: Adapted from *Oil & Gas Journal*, December 5, 2011, p. 34.

[a] Figures include partial interests in refineries not wholly owned by the company.

[b] Includes Shell's stakes in Motiva and its 50% stake in the Deer Park, TX, refinery.

[c] Consists of PdVSA's ownership of Citgo and its 50% stake in the ExxonMobil Chalmette, LA, refinery.

[d] Joint (50/50) venture between Shell and Saudi Aramco.

[e] Consists of 50% stake in Motiva.

Table 13–10. Companies in Western Europe with at least 200,000 b/cd refining capacity, January 1, 2012

Company	No. of refineries	Crude capacity,[a] thousand b/cd
1. Total	14	1,984
2. ExxonMobil	9	1,680
3. Royal Dutch Shell	9	1,197
4. Agip Petroli	10	876
5. BP	8	854
6. Repsol YPF	5	709
7. Turkish Petroleum Refineries	4	613
8. Petroplus International	5	521
9. Cepsa	3	427
10. Ineos Group Holdings	2	403
11. OMV	3	399
12. ERG Group	4	396
13. ConocoPhillips	3	350
14. Preem Raffinaderi	2	316
15. Hellenic Petroleum	3	313
16. Neste Oil	6	311[b]
17. Statoil	3	304
18. Galp Energia	2	304
19. Saras	1	300
20. PdVSA	8	295
21. Valero Energy	1	210

Source: Adapted from *Oil & Gas Journal*, December 5, 2011, p. 34.
Note: Western Europe includes Austria, Belgium, Denmark, Finland, France, Germany, Greece, Ireland, Italy, the Netherlands, Norway, Portugal, Spain, Sweden, Switzerland, Turkey, and the United Kingdom.
[a] Figures include partial interests in refineries not wholly owned by the company.
[b] Includes 50% stake in AB Nynas refineries.

Table 13–11. World's largest refineries, January 1, 2012

Company[a]	Location	Crude capacity, thousand b/cd
1. Paraguana Refining Center	Cardon/Judibana, Falcon, Venezuela	940
2. SK Corporation	Ulsan, South Korea	840
3. GS Caltex	Yeosu, South Korea	760
4. Reliance Petroleum[b]	Jamnagar, India	660
5. ExxonMobil Refining & Supply	Jurong/Pulau Ayer Chawan, Singapore	605
6. Reliance Industries[b]	Jamnagar, India	580
7. S-Oil	Onsan, South Korea	565
8. ExxonMobil Refining & Supply	Baytown, TX	561
9. Saudi Aramco	Ras Tanura, Saudi Arabia	550
10. Formosa Petrochemical	Mailiao, Taiwan	540
11. ExxonMobil Refining & Supply	Baton Rouge, LA	503
12. Hovensa	St. Croix, US Virgin Islands	500
13. Marathon Petroleum	Garyville, LA	490
14. Kuwait National Petroleum	Mina Al-Ahmadi, Kuwait	466
15. Shell Eastern Petroleum (Pte)	Pulau Bukom, Singapore	462
16. BP	Texas City, TX	451
17. Citgo Petroleum	Lake Charles, LA	440
18. Shell Nederland Raffinaderij	Pernis, the Netherlands	404
19. Sinopec	Zhenhai, China	403
20. Saudi Aramco	Rabigh, Saudi Arabia	400
21. Saudi Aramco–Mobil	Yanbu, Saudi Arabia	400

Source: Adapted from Oil & Gas Journal, December 5, 2011, p. 35.
[a] Only facilities with a capacity of at least 400,000 b/cd are shown.
[b] The two Reliance Petroleum facilities at Jamnagar together make that site the single largest refining operation in the world, with 1.24 million b/cd nominal crude processing capacity.

Refinery projects planned for 2010 and beyond

Oil & Gas Journal pointed to several significant refinery-related projects planned or under way in 2010.[10] China's plans are the most ambitious and include the following:

- A $3.4 billion, 200,000 b/cd refinery at Anning City, to process oil and gas to be imported from Myanmar beginning in 2013

- A $9 billion project (jointly owned by Kuwait National Petroleum and Sinopec) to build a 300,000 b/cd refinery on Donghai Island at Zhanjiang

- A $3.5 billion, 300,000 b/cd refinery and petrochemical complex planned for Tianjin, to be jointly owned by Rosneft and CNPC

- Doubling of capacity (to 200,000 b/cd) of PetroChina's refinery in Hebei province

- A new 200,000 b/cd refinery to be built by CNOOC in Dongying, Shandong Province

- Four petroleum refining bases—each with capacity of 200,000 b/cd—at Dushanzi, Urumqi, Karamay, and Kuqa, all in Xinjiang Uygur Autonomous Region

Activity planned in other nations includes:

- Commissioning of the 120,000 b/cd Cuddalore refinery in Tamil Nadu, India, to process crude into LPG, bitumen, and European-standard automobile fuels

- Expansion of two other refineries in India (at Mangalore and Vadinar) to boost capacity to 300,000 and 415,000 b/cd, respectively

- Tripling of the capacity of Bangladesh's only refinery (at Chittagong) to 100,000 b/cd, with assistance from a Saudi investment group

- Capacity expansion at Vietnam's only refinery (at Dung Quat) to 200,000 b/cd from 148,000 b/cd

- Investment by Indonesia's PT Pertamina and Kuwait National Petroleum of $9 billion in a joint venture, a 300,000 b/cd refinery in West Java

- Building of a complex south of Baku by Azerbaijan's state oil company to process 300,000 b/cd of crude and 1.4 tcf of gas per year, as well as to produce petrochemicals and electric power

- Investment by OPEC members in some 140 projects from 2012 to 2014, to expand refining capacity by 12 million b/d of crude and liquids, including a 400,000 b/d export refinery planned for Yanbu City, on Saudi Arabia's Red Sea coast
- Construction of a $4.8 billion refinery in southwestern Iran, starting in 2012
- Construction of a $4.6 billion refinery in northwestern Uganda, in the district of Hoima

In summary, the structures of both the upstream and the refining sectors of the global petroleum industry are complex and multifaceted. Dynamic interactions among a range of companies, governments, and other entities address a range of technical, economic, and public policy challenges to produce petroleum products for world markets.

References

1 PTCC. Mission statement. PTCC Web site: http://www.pttc.org/about.htm.

2 The Economic Contribution of the Onshore Independent Oil and Natural as Producers to the U.S. Economy: Final Report. April 2011. IHS Global Insight (USA), Lexington, MA.

3 Jessen, Rob. "Special Report: IOC challenge: Providing value beyond production." Feb. 2, 2009. *Oil & Gas Journal*, Vol. 107, Issue 5, p. 24. Tulsa, OK.

4 Pfeifer, Sylvia and Hook, Leslie. "Chinese demand for energy pumps up M&A share." Nov. 7, 2010. *Financial Times*. London.

5 Wood Mackenzie. "Chinese NOCs step up international expansion." June 15, 2010. Edinburgh, Scotland.

6 OPEC. Mission statement, from OPEC Web site: www.opec.org

7 OPEC. Membership criteria, from OPEC Web site: www.opec.org

8 *The Economist*. "Drill will: Disharmony at OPEC breeds uncertainty over the oil price." June 9, 2011.

9 True, W. R., and L. Kootungal. "Global capacity growth slows, but Asian refineries bustle." December 6, 2010. *Oil & Gas Journal*, Vol. 108, Issue 46, p. 50. Tulsa, OK.

10 Ibid.

14 Petroleum Trading

The global trading of petroleum is complex and dynamic. A myriad of buyers, sellers, brokers, traders, shippers, and other parties seek to discover and act on information about the supply/demand and prices of a wide range of energy commodities—crude oil and raw natural gas, as well as products ranging from jet fuel and industrial lubricants to gasoline, LNG, butane, and petrochemicals.

In this commercial arena, the air is filled with talk of strips and benchmarks; futures, hubs, and market baskets; spreads and swaps; contango and backwardation. Though arcane and somewhat intimidating to the casual observer, this specialized vocabulary describes with precision and clarity a range of situations, options, and actions. Detailed description of every aspect of petroleum trading—and its terminology—is beyond the scope of this book. Instead, the aim of this chapter is to provide a broad yet high-level perspective on this vital aspect of the petroleum industry, at the risk of occasional oversimplification. Readers interested in comprehensive information about petroleum trading can explore a range of books by other writers on this subject (e.g., see app. B).

The first section of this chapter presents information about the mechanisms and agencies that have evolved over the past 30 years or so to support and facilitate the trading of oil. Later sections address the natural gas market.

Oil Trading

Crude oil trading traditionally has involved oil from seven major regions:

- The North Sea
- The Russian-Caspian region
- The Mediterranean
- West Africa
- The Arab Gulf
- The Asia-Pacific region
- The Americas

Emerging sources include oil from Sakhalin Island off Russia's Pacific Coast and Nile Blend crude from the Sudan.

As noted in chapter 2, crude oils from different geologic sources and geographic regions can vary widely in composition, and different grades of crude have different relative values in the market. In general, lighter and sweeter crude (containing relatively little sulfur) is more valuable than heavy crude because it can be refined more easily into valuable products such as gasoline. This should be kept in mind as this discussion about trading oil and moving it to market unfolds.

Importantly, petroleum trading takes place in two largely discrete but occasionally intersecting worlds. The first is the wholesale, or *physical*, trade of actual volumes of oil; the second involves more-speculative, future-oriented, or *paper*, trading by use of specialized financial instruments.

Physical versus paper trading

In physical trading, one party agrees to purchase X barrels from a seller and then pays for and takes actual physical delivery of those X barrels (either right away or at a future date). Paper trading spells out the right or opportunity to buy or sell agreed-to quantities of oil by a specified date. Three kinds of instruments are commonly used: *futures contracts, options contracts*, and *swaps*. These instruments allow buyers and sellers to hedge against possible future swings in oil prices.

Futures and options are traded on an *exchange*, which is a marketplace where parties can buy or sell a commodity for delivery at some point in the future. The two major exchanges for oil trading are the NYMEX and the London subsidiary of Atlanta-based Intercontinental Exchange (ICE). Other principal exchanges are in Singapore and Tokyo. In addition, the Dubai Mercantile Exchange (DME) in 2007 launched its

Oman Crude Oil Futures Contract (OQD). DME reported almost 745 million barrels of crude traded in 2010 through the OQD.

With regard to exchange-based trading:

- A futures contract obligates the buyer to purchase (and the seller to deliver) a specific quantity of oil at an agreed-to price at a specified future date.

- The purchaser of an options contract has the right (but not the obligation) to buy or sell a specific amount of oil at a specified price, at any time during a specified period. If the option is not used (exercised), then it expires; no sale occurs and the money spent to buy the option is lost. An option trade can be conducted through an exchange or *over the counter* (OTC), meaning privately negotiated.

- In a swap agreement, a floating price for a specific amount of a specific type of oil is exchanged for a fixed price over a specified period. There is no transfer of physical oil; both parties settle their contractual obligations by a transfer of cash. The swap is an OTC transaction.

Futures and swaps do not require the investor to take or deliver the actual oil. In fact, in approximately 99% of all such trades, no actual oil is delivered.

In addition to the services offered by exchanges, several trading service providers offer oil (and gas) trading to individuals through Internet futures accounts. It is also possible to invest in exchange-traded funds that specialize in oil or natural gas futures contracts.

Pricing—a historical perspective

Decades ago, the industry established *benchmark crudes* to provide an index for general price movements. For example, in the early 1960s, following the formation of OPEC, the cartel set Arabian Light crude as a pricing benchmark.

In 1983, the NYMEX began trading crude oil futures contracts and chose West Texas Intermediate (WTI) crude, for delivery at Cushing, Oklahoma, as its benchmark. (This specification was later broadened to light sweet crude, which allowed traders more flexibility in fulfilling a contract if physical delivery of oil was required.) In 1988, the International Petroleum Exchange (now the ICE) chose Brent crude as its benchmark for futures contracts. Brent crude is actually a blend of light, low-sulfur oil from 20 different fields. Chapter 2 comments on the evolving battle between WTI and Brent crude for prominence in exchange-based trading.

Traders selling oil using a futures contract must be willing to deliver it if necessary. Thus, the prices determined on futures exchanges, such as the NYMEX and the ICE, correspond directly to prices that govern the physical flow of oil.

Disruption of markets caused by the Iranian Revolution of 1979–80 sent oil prices skyrocketing, and new suppliers entered the market to compete with OPEC. By 1985, Saudi production had plunged. In response, traders switched to an approach called *netback pricing*, in which crude prices were set by adding a normal refinery profit margin to the price of a collection, or *basket*, of refined products. Crude oil prices stabilized briefly, but in March of 1986, they fell to $10.42 per barrel on the NYMEX.

In 1986, OPEC established a production allocation program, aiming to manage oil price swings by setting production quotas for its members. As discussed previously (in chap. 13), the quota system has been quite difficult to enforce, and oil-price volatility continues to this day. (For example, in 2008, crude prices ranged as high as $145 and as low as $36 per barrel.)

Oil pricing today

Wholesale trade in actual oil is conducted through either *term* or *spot* supply contracts. The majority of oil is traded today using the former, in which a seller agrees to supply a specific quantity of oil to the buyer for an agreed-to price at one or more future dates. In contrast, spot supply contracts are for delivery of a quantity of oil at a specific location as soon as possible (often, within a day or two).

On the spot market, a wide range of commodities are bought and sold for cash and delivered immediately, with payment literally on the spot. The prices of commodities (including oil and gas) are subject to extreme fluctuations—for example, caused by natural disasters, political developments, weather events, and changes in estimates of supply or demand. Nonetheless, the spot market does provide signals about expected near-term prices.

Spot contracts are arranged on the international market by traders operating around the world, at firms based in London, New York, and Singapore, as well as Geneva and Baar, in Switzerland; Amsterdam and Rotterdam, in the Netherlands; and even Westport, Connecticut. Spot supplies are sold to the highest bidder willing to take delivery quickly.

When sold for immediate delivery, the price of oil is called the *spot price* or *cash price*; when sold for delivery at a specified future date, the price is called a *forward price*.

Until the 1980s, oil prices established in term supply contracts were fixed by major oil companies, unilaterally or by negotiation, and would not change for weeks, months, or even years at a time, based on the contract terms. Today, however, almost all pricing for term supply contracts is based on benchmark pricing. Although it may seem circular or counterintuitive, benchmark prices are based in part on spot prices.

Benchmark oil prices (also called *price markers*) are set at the close of each business day on the basis of two sources of data: (1) exchanges (e.g., the NYMEX and the ICE) on which oil futures contracts are traded; and (2) a group of well-regarded oil trade journals (of which *Platts Oilgram* and *Petroleum Argus* are the two most widely used). The journals assess spot-market price information (sent by traders and brokers) about hundreds of grades of oil and related products at locations around the world (e.g., New York Harbor jet fuel), based on physical spot-market trading.

The following are well-known benchmark oil prices:
- NYMEX WTI (or Light Sweet) Crude
- Light Louisiana Sweet Crude
- NY Harbor Jet Fuel
- ICE Brent Crude
- Dated Brent Crude
- Rotterdam Barges Fuel Oil
- Dubai Crude
- Singapore Gas Oil

Also notable is OPEC's collection of price data on a basket of crude oils (and blends) and their calculation of average prices for these oil streams to develop an OPEC Reference Basket price to monitor world oil market conditions. As of October 2011, the OPEC Reference Basket was made up of the following:
- Saharan Blend (Algeria)
- Girassol (Angola)
- Oriente (Ecuador)
- Iran Heavy (Iran)
- Basra Light (Iraq)
- Kuwait Export (Kuwait)

- Es Sider (Libya)
- Bonny Light (Nigeria)
- Qatar Marine (Qatar)
- Arab Light (Saudi Arabia)
- Murban (United Arab Emirates)
- Merey (Venezuela)

The prices most commonly referenced in term oil contracts reflect recent activity in major trading locations such as Singapore, the Netherlands, New York Harbor, Dubai, the U.S. Gulf Coast, and Los Angeles. Prices established in contracts for nonbenchmark grades of oil are set at a premium or discount to one or more of the benchmark grades, based on quality differences, transportation costs, and taxes for the oils being compared. Complex formulas are sometimes used to determine the exact premium or discount each day or each month.

Pricing along the delivery chain

Prices for physical oil are often quoted in relation to a certain point in the delivery chain. The crude oil price, established early in the process, accounts for most of the price charged for the final product (e.g., gasoline), but each step in the refining and transportation chain adds to that final price.

Not all grades of oil follow the delivery sequence shown below, or they may move through the steps noted in a different order; nonetheless, the details are informative:

- *Wellhead price*: The price of crude at the point of production
- *Cargo price*: The price of crude or product on a large oceangoing tanker
- *Refinery gate price*: The price of crude entering a refinery or of finished products leaving a refinery
- *Barge price*: The price of oil on a barge vessel typically used on a river or close to shore, with a typical capacity of 8,000–50,000 barrels
- *Pipeline price*: The price of oil at a specified point in a pipeline
- *Rack price*: The price of oil at a wholesale terminal from which tank trucks take oil to be carried to filling stations, homes, and businesses

- *Dealer-tank-wagon price*: The wholesale price of oil delivered to retail outlets, such as gasoline stations
- *Retail price*: The price of oil to the end user, which typically includes a retail tax and a markup imposed by the retail station owner.

Posted prices

Another mechanism that affects global oil prices is the *posting* of oil prices. Refinery owners use this approach to specify (post) an *offer price* each day that they are willing to pay for various crude oil grades. Posted prices are for oil delivered at a specified point and of a specified quality—oil produced from a particular field or a blend of common crudes. Knowing the source and quality (e.g., API gravity) of the crude it buys allows a refiner to better estimate how much of various refined products can be produced from the crude feedstock. The price that refiners are willing to pay will shape relative crude prices for particular fields and for oils of different quality. This posted price procedure was dominant in North American oil markets until the rise of the exchanges (e.g., the NYMEX and the ICE) for trading oil contract futures and options, as described earlier in this chapter.

Hubs

Much of the market-related activities associated with price discovery and transfer of ownership for crude oil (and natural gas) take place at *hubs*. In general terms, a hub is a point at which pipelines come together. A hub on the coastline is called a *marine terminal*; when it draws together output from several oil or gas wells, it is called a *gathering station*. Substantial storage capacity is also typically put into place at or near hubs.

There are several major crude oil pipeline hubs in the United States. Perhaps the best known is at Cushing, Oklahoma, where the NYMEX crude oil futures contract is physically deliverable (if necessary). Other primary hubs and marine terminals for oil-price discovery in the United States are:

- New York Harbor
- The Gulf Coast (along Texas and Louisiana)
- Group Three (Tulsa, Oklahoma)
- Chicago
- Los Angeles

Important international locations for oil price discovery include the following pipeline hubs or marine terminals:

- ARA (Amsterdam, Rotterdam, and Antwerp), in the Netherlands/Belgium

- Genoa/Laverna, on the Italian/French Mediterranean coast

- Singapore

- The Arab Gulf—particularly Ras Tanura (Saudi Arabia) and Dubai (United Arab Emirates)

There are also market hubs for *petroleum products* at New York Harbor, the Houston Ship Channel, Los Angeles, and Group Three. NGL market centers are located at Mt. Belvieu, Texas; at Conway, Kansas; at Sarnia, Ontario; and in the Los Angeles Basin.

Transportation services

An important component of the midstream petroleum sector is the transport of crude oil and petroleum products by pipeline, tanker ship, railcar, and tank truck. Some large producers own and operate their own transportation equipment, but most use independent transportation operators for some or all oil movements.

Some of the larger firms offering oil tanker ship services are shown in table 14–1. (For information about major U.S. oil pipelines, see chap. 9.)

Table 14–1. Oil tanker ship companies

Company	Location
A/S Steamship Company Torm	Copenhagen, Denmark
Blue Fin International Shipping	Dubai, UAE
Crowley Marine	Jacksonville, FL
Frontline	Hamilton, Bermuda
General Maritime	New York, NY
Kirby	Houston, TX
Nippon Yusen Kabushiki Kaisha (NYK Line)	Tokyo, Japan
OMI	Stamford, CT
Overseas Shipholding Group	New York, NY
Seacor Holdings	Ft. Lauderdale, FL
Ship Finance International	Hamilton, Bermuda
Teekay Shipping	Vancouver, BC
US Shipping Partners	Edison, NJ

Source: Downey.

Although many large companies in the oil industry own their own vessels, a big part of the global tanker fleet is owned by individuals and corporations with no physical crude oil production or oil-refining capacity. Business people in Greece and Italy, for example, are prominent shipowners.

Individual owners usually use brokers to find cargoes, using a process called *fixing a charter*. First, the party initiating the charter (e.g., a refiner or oil trader desiring transport service) contacts a broker or a shipowner. The initiator is ready to specify the product or crude to be carried, the size of vessel needed, the route, the time period for delivery, and possibly the ports of loading and unloading at either end of the route. Tanker vessels are sometimes chartered as stationary floating storage tanks—for periods when storage costs onshore are very high or when onshore tanks are full.

The charterer and owner agree to a price for the transaction and provisionally book the vessel. In most cases, the parties refer to a document published annually by the not-for-profit Worldscale Association that lists nominal voyage costs (in U.S. dollars per metric ton of cargo). The book contains rates for travel between regions (e.g., U.S. Gulf Coast and the Baltic, Caribbean, and Black seas), as well as between individual ports. It is the responsibility of the charterer to ensure that the vessel meets all requirements (e.g., has a double hull) pertinent to ports at which the vessel may load and unload.

When all terms are agreed to and all issues are in good order, the charterer and vessel owner select one of three types of charter agreement:

- *Spot charter* (also called *voyage charter*)
- *Contract of affreightment* (COA)
- *Period charter*

The spot charter is for use of a specific vessel in the immediate future to move cargo between specified loading and discharge ports; it is a "one-shot" arrangement. The major difference between a spot charter and a COA charter is that the COA agreement does not stipulate the use of a specific vessel.

Whereas spot and COA charters refer to Worldscale Association data for pricing, period charters establish an agreed-on day rate (in U.S. dollars per day) for a particular vessel. There are several types of period charter; however, their description is beyond the scope of this book.

Standard trade terms used in global shipping have been developed by the International Chamber of Commerce. Among the most common terms used in defining tanker freight charges are five methods of delivery pricing:

- *FOB* (free on board): The seller is no longer responsible for the oil once it has been loaded onto the ship; the buyer assumes all risks and costs of ownership from that point.

- *CFR* (cost and freight; also called *C+F*): The seller pays all costs incurred to bring the oil to the delivery port, but the buyer is responsible for insuring the cargo while in transit. Once the cargo leaves the ship, all further costs accrue to the buyer.

- *CIF* (cost, insurance, and freight): The same as CFR, except that the seller also is responsible for insuring the cargo while in transit.

- *Landed cost*: The same as CIF, plus the seller pays the import duties at the delivery port.

- *DDP* (delivered duty paid): The seller is required to deliver the oil to the location specified by the buyer with all insurance, freight costs, and duties paid; in short, the DDP price includes all costs.

Sale and movement of refined products

After crude oil is converted into refined products, the products are transported and sold to end users around the United States. The collection and dissemination of data regarding this aspect of the petroleum industry is conducted using five *Petroleum Administration for Defense Districts* (PADDs). They are named after a government agency, the Petroleum Administration for Defense, established in 1950 in conjunction with the Korean War effort; even though the agency was eliminated in 1954, the name is still used. These districts (see tables 14–2 and 14–3) were created during World War II to help organize the allocation of fuels derived from petroleum products.

In many cases, refined products move from a PADD that has excess supply to another that needs more products. For example, PADD III refineries supply a substantial volume of products to the Midwest (PADD II) and the East Coast (PADD I). As described in chapter 5, refined products are moved to market centers by using product pipelines, as well as tanker vessels and tank trucks.

Table 14–2. States in each PADD

PADD	States
I	Maine, Vermont, New Hampshire, Massachusetts, New York, Pennsylvania, Rhode Island, Connecticut, New Jersey, Delaware, Maryland, West Virginia, Virginia, North Carolina, South Carolina, Georgia, Florida
II	North Dakota, South Dakota, Nebraska, Kansas, Oklahoma, Minnesota, Iowa, Missouri, Wisconsin, Illinois, Michigan, Indiana, Ohio, Kentucky, Tennessee
III	New Mexico, Texas, Arkansas, Louisiana, Mississippi, Alabama
IV	Montana, Idaho, Wyoming, Utah, Colorado
V	Washington, Oregon, California, Nevada, Arizona

Source. EIA.

Table 14–3. Consumption of petroleum products by PADD in 2010

Product(s)	Consumption by PADD, thousand b/d					
	I	II	III	IV	V	Overall
Still gas	39	127	350	23	131	670
Gasoline	3,248	2,531	3,585	292	1,544	11,200
Jet fuel	544	254	158	51	471	1,478
Distillate fuel oil	1,169	1,159	796	177	492	3,793
Residual fuel oil	259	24	103	8	157	551
Asphalt and road oil	85	149	67	33	27	361
Petroleum coke	31	141	148	22	34	376
Petrochemical feedstock	18	45	404	0	1	468
Lubricants	33	28	63	0	8	132
Waxes	1	1	5	0	1	8

Source: EIA.

Petroleum product marketers

A few words are warranted here about the role of petroleum marketers in delivering a range of refined products (gasoline chief among them) to U.S. consumers. Any individual or company that takes possession of refined petroleum products for the purpose of reselling those products is considered a petroleum marketer; this definition spans a wide range of commercial businesses, both wholesale and retail. A petroleum marketer often owns gasoline stations, convenience stores, heating oil businesses, truck stops, lubricant warehouses, petroleum trucking companies, and bulk storage facilities.

Independent marketers are companies that purchase refined products (from major or independent refineries) and sell them at retail outlets. This classification includes firms whose primary business is gasoline sales (e.g., Speedway and QuikTrip), as well as large retailers (e.g., Costco and Walmart) who sell gasoline in addition to other merchandise. Two trade groups represent this group: the Petroleum Marketers Association of America and the Society of Independent Gasoline Marketers of America.

Gasoline distribution and marketing

After typically brief storage at a supply terminal, refined gasoline is distributed to service stations, airports, and other outlets by tank trucks. Companies mix proprietary additives into gasoline at these facilities.

Distributors are companies without refining operations that own and operate their own gas stations. Distributors may also supply gasoline to dealers who are independent gasoline stations with no distribution capabilities. More than 114,000 gasoline stations delivered transportation fuels to motorists in 2008 throughout the United States, according to census data.

Gasoline is sold at branded and unbranded stations. Branded stations include company-owned facilities, distributors, and dealers. Company-owned stations—which are owned and operated by a petroleum refining company (e.g., ExxonMobil and Chevron)—usually focus more on fuel sales than do operators of convenience stores. Branded operations allow distributors and dealers to market gasoline under a recognized brand. Unbranded stations can operate as either retail or commercial entities.

Natural Gas Trading

A better understanding and appreciation of the dynamics at play in today's U.S. natural gas market can be achieved by reviewing, even briefly, the dramatic change that took place in the industry about 25 years ago.

U.S. historical perspective

Until about the mid-1980s, pipelines and LDCs operated under the direction of federal and state agencies that controlled the price of natural gas at the wellhead and that regulated the rates that customers paid for gas. Pipelines bought gas from producers, moved it to market, and then resold the gas to LDCs.

Gas producers signed long-term contracts (typically for 20 years) with pipelines, promising to supply gas, and the pipelines agreed to pay for that gas regardless of whether they could sell it. As a result, in the late 1970s, more and more pipelines found themselves saddled with extra gas they could not sell but for which they still had to pay high prices.

In response, federal authorities deregulated gas sales through the Natural Gas Policy Act of 1978. The Federal Energy Regulatory Commission, which oversees interstate pipelines, issued a series of orders—FERC Order 436 (in 1985), 500 (in 1987), and 636 (in 1993)—that permitted LDCs and large gas customers to buy gas directly, not only from producers but also from marketers and brokers. Pipelines also were required to provide comparable transportation service for natural gas, regardless of where or from whom the customer had bought it.

In short, pipeline companies were transformed from merchants (who bought and then resold gas) into transporters. In response to this government action, hundreds of new players entered the arena as gas marketing companies or brokers. Some were created as subsidiaries of pipelines, producers, and LDCs; others were independent; but all aggregated gas from several sources into large volumes for sale to LDCs and large gas users (e.g., industrial firms and power plants).

Physical delivery of natural gas

Deregulation and restructuring also sharply altered the terms of gas purchase contracts. By the late 1980s, the familiar 20-year contract was replaced by shorter-term arrangements (typically, of 30 days' duration). Since then, creative purchase agreements have evolved that offer a mix of short-, medium-, and long-term contract provisions tailored to meet the needs of the customer.

Pipeline companies have had to adapt to the increased number of players and the variety of transactions and contract arrangements that characterize the modern natural gas market. For example, a gas customer may have a two-year contract with a supplier to buy a certain volume of gas over that period. Even so, on each day, the customer also is contractually required to submit a *nomination* to the pipeline company, specifying how much of that gas to deliver in the next 24 hours and where to deliver it. Using this information, the pipeline company takes action to ensure that gas flow through the delivery system will be properly balanced.

In some cases, a customer may opt for lower-cost, *interruptible* delivery service. This gives the pipeline operator flexibility to deliver gas first to *firm* customers, after which the remaining capacity is made available to interruptible customers.

Hubs

Creation of natural gas market centers and hubs began in the late 1980s in the United States. These evolved as an outgrowth of natural gas market restructuring and the execution of a number of FERC Orders (described earlier) that forced interstate gas pipeline companies to become primarily natural gas transporters.

The most prominent U.S. market center for natural gas is the Henry Hub, the location of a gas plant on the natural gas pipeline system in Erath, Louisiana (approximately 15 miles south of Lafayette and 60 miles southwest of Baton Rouge). The Henry Hub, which has been the pricing point since 1992 for natural gas futures contracts traded on the NYMEX, interconnects with nine interstate and four intrastate pipelines. North American unregulated wellhead and burner-tip natural gas prices are closely correlated to those set at Henry Hub.

European gas hubs

Market hubs for natural gas trading also arose in Europe over the past several decades, where substantial volumes of gas flow between and within many countries. In 2010, for example, physical spot trading on the seven leading continental hubs reached 4.94 tcf (140 billion cubic meters). Traded (non–spot-market) volumes were about three times that figure.

The seven major European spot-market gas-trading hubs are:
- Zeebrugge, Belgium
- TTF (Title Transfer Facility), the Netherlands
- NGC (NetConnect Germany), Germany
- Gaspool, Germany
- PEGs (Gas Exchange Points) North and South, France
- PSV (Punto di Scambio Virtuale), Italy
- CEGH (Central European Gas Hub), Austria

Market centers or hubs offer two key services: transportation between (and interconnection with) other pipelines; and the physical coverage of the balancing needs associated with short-term gas receipt or delivery. Most will conduct transfer-of-title transactions between parties that buy, sell, or move their natural gas through the center, and many also provide other services that help expedite and improve the overall gas transportation process. These services include Internet-based access to natural gas trading platforms and capacity release programs. At the end of 2008, there were a total of 24 operational market centers in the United States and nine in Canada.

Financial gas market

Deregulation of wellhead prices opened the door to the trading of natural gas as a commodity on financial exchanges, as well as the straightforward physical transportation and delivery of actual quantities of gas. The players in this part of the arena are traders and brokers, who, unlike marketers, never actually take ownership of the gas. They often make short-term deals—with terms ranging from a few months to as short as one day—seeking to capitalize on anticipated or actual changes in gas prices in different locations. These market participants use a variety of financial tools to spread risk, stabilize prices, and enable an owner of a quantity of gas to quickly convert that asset into cash if desired. The information provided earlier in this chapter about physical versus paper trading, the spot market, and the use of futures contracts, options contracts, and swaps in oil trading applies equally in the natural gas market.

LDC role in gas marketing

Reflecting developments in the industrial and large-volume gas sales market, LDCs have offered customer-choice programs—also known as *retail unbundling programs*—since 1998 to residential and commercial customers. Under these programs, consumers can select their own gas provider. When a customer buys gas from a provider other than the LDC, the LDC still delivers it, even though the LDC is not involved in the sales transaction. In essence, the customers who switch to another provider become transportation-service customers for the LDC.

Despite this change in the traditional LDC/customer relationship, LDCs continue to promote efficient and safe use of the natural gas they deliver and to offer technical support to engineers, architects, builders, and commercial and industrial customers seeking to buy or install equipment. Many also offer energy-conservation programs and energy audits to customers.

Few LDCs directly sell gas-fueled equipment. Nevertheless, some do participate in cooperative advertising campaigns with appliance manufacturers and local dealers.

LNG trading

Chapter 10 described the equipment and processes used to convert wellhead natural gas to supercold liquid form (LNG) for transport to customers. This section briefly describes some of the major financial and business aspects of LNG production and shipment.

The typical LNG project is both large and complex, involving not only the upstream operation of liquefaction plants and export facilities (for producing, temporarily storing, and then loading LNG onto specially designed tanker ships) but also the chartering and travel of those ships to import facilities (at which the LNG is regasified and introduced into pipelines for distribution). Each LNG project has its own unique characteristics and organizational structure, determined by legal and financial issues, ownership of the gas itself, political goals of the exporting country, and decisions about responsibility for various aspects of the value chain, described previously.

Gas producers—sometimes including (or solely) an NOC (see chap. 13)—tend to be involved chiefly in the gas production and liquefaction stages. They also may want control of the LNG shipping process. However, LNG customers are said to be expressing more interest in participating in those upstream stages, to increase security of supply and to participate as investors in the more profitable parts of the LNG chain. They are increasingly acquiring control of LNG shipping and making purchases more often on an FOB basis. Examples include Osaka Gas, Tokyo Gas, Kogas, Union Fenosa, Gas Natural, and Gaz de France.

At the same time, some LNG producers are moving downstream, seeking to secure market outlets through ownership or purchase of access rights to LNG import terminals in competitive markets, thus gaining more control over ultimate sale of the gas. Examples include BP, Shell, Chevron, and ExxonMobil (all IOCs), as well as the NOCs Statoil, Petronas, and CNOOC.

LNG sales have traditionally been on a long-term basis (20 years or more), through purchase contracts with large LDCs or electric power generation companies that are state-owned or have a monopoly service territory. This long-term characteristic of LNG contracts has given shareholders and lenders the confidence to make substantial commitments of capital to LNG projects.

Pricing of LNG has until recently been dependent on the buyer's market conditions. For a buyer such as a large utility with no pipeline-gas supply (e.g., in Korea and Japan), LNG prices have been set primarily by reference to crude oil—subject to approval by governmental authorities. However, in cases where security of supply has become as important as price, LNG pricing can become more competitive. In Europe, for example, LNG competes with gas delivered by pipeline from Russia and elsewhere; there, LNG pricing tends to follow pipeline pricing, with both generally linked to oil or oil products when they are supplied under long-term contracts. In the United States, LNG is generally priced against the Henry Hub index, often with adjustments for differences in delivery location.

At the same time, the rapid development of an LNG spot market—coupled with growing availability of shipping not committed to long-term contracts—have led to increasing globalization of spot LNG prices. Future expansion of LNG markets will depend on where new gas resources are discovered and developed, and on the relative cost of transporting that gas by pipeline compared to the LNG option. Still, there appears to be growing acceptance of the view that transportation as LNG can be cost competitive, based on evidence of impressive cost reductions over the past decade or so in LNG production and delivery infrastructure. For example, Qatar has put into place very large liquefaction facilities whose economies of scale make it possible to deliver large quantities of gas to the United States at costs competitive with many indigenous North American gas resources. It remains to be seen, however, where future domestic North American gas will come from—the Alaskan North Slope, for example, or the Canadian Arctic, or the burgeoning shale gas formations of the Lower 48—and how the eventual market price for that gas will compare to imported LNG.

The United States may even begin to export LNG as domestic production increases. Two proposals in late 2010 (one by Texas-based Cheniere Energy, the other by Australian bank Macquarie in partnership with Freeport LNG) would liquefy for export as much as 3.4 bcf per day of domestic U.S. natural gas.

15 Emerging Challenges for the Petroleum Industry

Over the next several decades, the petroleum industry faces a future likely to offer not only significant opportunity but also numerous tests of its ability to adapt and innovate. Disruptive yet potentially energizing trends and issues on the near and immediate horizons include the following:

- IOCs, finding it much harder to access new oil reserves, are having to turn to new areas, such as unconventional oil, deepwater basins, and unconventional gas.

- In 2009, the non-OECD economies were responsible for all of the growth in global energy demand. Since 2000, their share of demand growth has averaged more than 90%, according to BP.

- The influence of OPEC is rising once more, and one member, Iraq, is poised to again become a major global oil supplier in coming years.

- Successful development of natural gas from shale has triggered a reassessment of that fuel's role, both in the United States and around the world. Several IOCs are studying possible large-scale shale gas production in China.

- The United Arab Emirates is building the first nuclear power plant in the Arab Gulf region—in a country that is one of the top 10 holders of oil reserves in the world. The 1,400-megawatt power station at Braka, in Abu Dhabi—the first of four planned units—is slated to start up in 2017.

- China is simultaneously pursuing oil and gas resources around the world, expanding its SPR, developing renewable energy resources, pushing hard to become more energy efficient, and growing its nuclear power generation capacity.

- Since about 2006, coal has been the fastest-growing primary energy source, used chiefly for power generation in industrializing economies, notably in China.

Projections by a range of government agencies and major industry players make it clear that oil and natural gas will continue to play an important role in the global energy marketplace in the next quarter-century. EIA, for example, judges that crude oil and associated liquids will provide approximately 32% of total global marketed energy use in 2035, with natural gas providing 22%.[1] (If coal is added to the calculation, fossil fuels will provide just over 80% of total world marketed energy use that year, compared to 84.1% in 2010.)

Of immediate concern, however—to a wide range of companies, not just the petroleum industry—is the lingering burden of a global economic recession that began in 2008, continued into 2009, and was still affecting energy use in 2012 in a number of countries (chiefly the industrialized OECD nations). Recovery has been both uneven and painfully slow. Developing non-OECD Asian economies have led the way, and many are well on the road back to economic health. The recession in the United States may be formally over, but housing and employment data indicate that economic growth is still at anemic levels in early 2012. Recovery is also lagging in Japan, and several European nations (particularly Greece) continue to struggle with serious economic issues.

Looking further ahead, the EIA analysis noted above offers cautious optimism. It assumes that by 2015, most nations of the world will have resumed the expected rates of long-term growth they saw before the recession. For the period from 2007 to 2035, world gross domestic product is expected to rise by an average of 3.2% per year, with non-OECD economies averaging 4.4% and OECD economies 2.0% per year. Against that backdrop, EIA projects total world consumption of marketed energy increases to grow by 49% over the 2007–2035 time frame, with the largest increase occurring (as noted above) in non-OECD economies.[2]

More-detailed or longer-term prediction of events and trends affecting the petroleum industry is difficult at best. However, it is realistic to expect that the industry will be pushed hard to marshal its resources and skills in new ways as it works to develop oil and gas resources to meet the needs of future energy consumers. The remaining pages of this chapter describe some of the challenges expected to confront the petroleum industry in the decades ahead. The intent here is not to presume knowledge of the answers; rather, the objective is to set forth some of the major issues facing the petroleum industry and to suggest actions likely to be required in a dynamic global energy marketplace.

Continued Technology Development

Leveraging its own resources with those of government and other entities, the petroleum industry will need to continue to develop and apply (as it has in the recent past) new technologies to maintain competitive advantage. Anticipated areas of focus include new approaches to:

- Increasing the success rate for finding new hydrocarbon resources and identifying those most promising for development
- Optimizing production from both new and legacy wells
- Controlling operating costs and improving efficiency through expanded use of automation, computer-based technology, advanced imaging systems, and new materials
- Complying with evolving requirements for environmentally acceptable operations.

The first two items of the above list are expected to be of particular importance in the exploitation of such truly challenging frontier resources as methane hydrate or those likely to be found in the Arctic or far below the seafloor, in ultra-deepwater.

Competition from Nonfossil Energy Sources

The EIA projection cited above notes that renewable energy resources accounted for just over 10% of total global energy consumption in 2010. However, this figure is expected to jump to 13.5% by 2035.[3] (For more insights into this trend, see the below section "The Chinese Challenge.")

Of particular interest for the petroleum industry will likely be the growing use of plant and waste materials to produce liquid and gaseous fuels that could compete with oil and gas for market share. From cellulosic ethanol to blue-green algae to advanced digester design, there may be opportunities for the petroleum industry to participate in bringing fuel from renewable resources into the global energy mix.

On a different front, support for nuclear energy as a source for electric power generation is weakening in some countries in the wake of the March 2011 earthquake and tsunami that seriously damaged nuclear power reactors in Fukushima Prefecture, Japan. Therefore, replacement of some nuclear generating capacity with increasing supplies of natural gas may become attractive.

Climate-Change Policy

With a view to the big picture, decisions by governments on whether, how, and when to deal with global climate change would likely have the most significant impact on use of fossil fuels (and renewable energy resources) around the world. The year 2012 marks the end of the first commitment period under the Kyoto Protocol, after which a new international framework will need to be negotiated and ratified. The Kyoto Protocol was signed in 1997 by 84 nations, launching international efforts to control greenhouse-gas emissions and combat global climate change.

Agreements reached at the U.N. climate conference in Cancun, Mexico, in December 2010 gave participating nations another year to decide whether to extend the 1997 Kyoto Protocol. Other provisions established a $100 billion fund to help poor countries adapt to climate change and created new methods to transfer clean energy technology.

Just before the Cancun meeting, Paris-based IEA issued its annual *World Energy Outlook*, which urged governments to implement their pledges (made in Copenhagen in late 2009) to fight climate change and cut fossil fuel subsidies. The agency warned that failure to do so would drive oil prices significantly higher. IEA noted that the big differences in price and demand outcomes under the different scenarios it evaluated would complicate investment planning by oil and gas companies, as well as policy decisions by OPEC.

At the next UN climate change conference—in Durban, South Africa, November 28–December 11, 2011—negotiators finally agreed at the eleventh hour to be part of a legally binding treaty whose terms are to be defined by 2015 and which would take effect in 2020.

Industry Structure

The continued viability of the business model of the vertically integrated major oil company has come into question following (*a*) the explosion at BP's Macondo well in the Gulf of Mexico, in April 2010, and (*b*) the rather lackluster 2005–10 financial performance of the half-dozen supermajor oil companies (described in chap. 13). Some analysts have suggested that spinning off refining assets (as both Marathon and ConocoPhillips did in mid-2011) might make it easier to partner with state-controlled companies or to participate in joint ventures. Others have asserted that separate entities focused on exploration and on production/refining might make more sense. The principal counterargument is that only very large companies can handle the rising costs and risks of exploration as stiffer regulations are developed in response to the Macondo accident.

On a related note, some observers suggest there may be a shift coming in the willingness of large and small companies to work together. Small companies may avoid major projects because of rising costs and potential liabilities; big firms may conclude that small companies don't have the financial resources to adequately share the risk.

Industry Operations

Workforce demographics

One of the biggest challenges facing the U.S. petroleum industry is to build and develop its next-generation workforce to replace (*a*) the 500,000 industry professionals lost in staff reductions from 1982 to 2000 and (*b*) the large numbers of baby boomers expected to retire in the next few years. It will be critical that new workers and leaders have not only the required technical training but also the communication and policy-analysis skills to plan and manage large and complex energy development projects.[4]

An analysis published in September 2011 by *Rigzone* described several initiatives that can help address the situation. These include competency development for new hires in critical jobs (e.g., reservoir engineering), best-practices collaboration, tailored education initiatives, and increased use of mentoring.[5]

Role of natural gas

Companies worldwide will likely continue to reassess the relative focus they give to oil and natural gas development. Firms will determine how the return of natural gas to a dominant position in the U.S. and global energy mix—driven by dramatic production from gas-bearing shales—factors into strategic corporate decisions.

A related issue is LNG's role in the U.S. and world energy markets. Continuing advances in technology for gasification, storage, regasification, and even ocean shipping could bring down the cost of LNG. Further, as exploration companies pursue ever more remote resources, LNG might take on a more dominant role in bringing that gas to market. A related issue emerging in the United States is the potential for increased exporting of LNG, in view of burgeoning domestic production.

Crude oil quality

For the projected global increase in demand for crude oil to be met, some analysts speculate that a portion of tomorrow's supplies will come from a range of new sources, chief among them Brazil, West Africa, Iraq, Canada, Kazakhstan, and Russia. An analysis in 2010 by Purvin & Gertz suggested that global crude oil is likely to become slightly heavier and more sour (i.e., contain more sulfur) over the next several years.[6] The largest net change in crude oil quality during the next 10 years likely will be in North America as Canadian oil, primarily heavy and sour, displaces traditionally sweeter crude from the United States. These expected changes in quality will affect technical refinery operations and will have an impact on feedstock pricing in various markets around the world.

Environmental and safety compliance

A major topic of debate in 2011 and 2012 has been the environmental impact of the hydraulic fracturing process used to extract natural gas from shale formations. Concerns center on possible contamination of groundwater either by (a) the chemicals pumped at high pressure into the ground as part of the fracing fluid or (b) the improper handling of water and fluid brought back to the surface for treatment and disposal. There have also been reports that methane gas (possibly released during fracing) migrates upward through the ground and dissolves into drinking water drawn from aquifers not far below the surface.

Two other recent significant events were:

- The loss of life and highly visible environmental damage caused by the April 2010 blowout of BP's deepwater Macondo well in the Gulf of Mexico

- The September 2010 gas pipeline explosion in San Bruno, California (near San Francisco), that killed eight people and destroyed 37 homes

All these incidents underscore how vital it is that the petroleum industry not only comply with regulations but also help all stakeholders understand industry actions taken to ensure public well being. This diligence applies as well to other operations, from the siting of LNG terminals to the construction and operation of refineries and other major energy-system components.

Security

Terrorist attacks on the United States and other nations since 2000 have heightened the urgency of protecting of the facilities, assets, and staff of the global petroleum industry. Security measures now range from patrolling sea-lanes to prevent pirate attacks on tanker vessels, to strengthening computer and security systems from cyber attack, to carefully managing intellectual property and proprietary information.

Maritime piracy costs the global economy $12 billion per year, according to researchers. Particularly troublesome are the areas off the Gulf of Aden and along the eastern coast of Africa, extending southward from Somalia, through which large numbers of oil tankers move.

Average ransom payments (for hijacked ships of all kinds) jumped to $5.4 million in 2010, compared to just $150,000 in 2005, according to a December 2010 analysis by Colorado-based One Earth Future Foundation. The largest ransoms have been reported for release of oil supertankers; in November 2010, a record $9.5 million was paid for the release of the South Korean VLCC *Samho Dream*, captured while carrying a cargo of Iraqi oil headed for the United States.[7]

Also of concern are cyber attacks. The FBI has reported that hackers based in China had stolen information from at least five U.S. oil and gas companies during 2010, including data about the configuration of key equipment.

The Chinese Challenge

As alluded to at the start of this chapter, all other players on the global petroleum stage will need to stay abreast of the multifaceted energy strategy being pursued by China. News agencies pay close attention China's aggressive global search—from Latin America to Africa—for oil and gas (and business deals to secure them) to meet growing domestic demand. China was second only to the United States as a net importer of oil in 2011. Moreover, China has been active on the international scene (see chap. 13), buying minority stakes in challenging deepwater blocks overseas. The country's first deepwater drilling platform was under construction in 2011 in Shanghai.

China is also aggressively pursuing energy conservation and development of non–fossil fuel (nonfossil) resources, as ways to achieve energy security and to reduce combustion-related emissions. For example, China aims to cut carbon emissions by 40%–45% before 2020 and meet 15% of domestic energy needs using nonfossil resources. Two parts of that program will be the expanded use of nuclear energy for power generation and a $750 billion stimulus program for renewable energy development. The latter includes not just wind, solar, and hydropower but also unconventional gas (e.g., from coal seams or shale) and clean coal.

Other International Interactions

A reversal of roles is occurring between NOCs in resource-rich countries and their traditional IOC rivals. In the past, the IOCs offered capital and technology to nations seeking to develop their hydrocarbon resources, under profitable contract arrangements. Now, however, many NOCs have access to funds from other sources (including their own treasuries) and can seek help instead from major global service companies.

The supermajors will have to adapt to survive. As described in chapter 13, it appears that IOCs will need to work harder to convince NOCs and host governments that they can provide value beyond just finding and producing oil and gas.

Also on the international front, the importance of OPEC is expected to grow, according to BP's *Energy Outlook 2030*, published in January 2011. The cartel's share of global production is expected to rise from 40% in 2010 to 46% by 2030, a level not seen since 1977.[8] More details about the structure and intraorganizational dynamics of OPEC are presented in chapter 13.

Peak Oil Production

An ongoing debate among petroleum industry observers concerns peak oil production. At issue is whether the point of maximum annual production of crude oil has already been passed or still lies ahead. The question is posed for both total global production and individual countries.

Although the complexities of the issue and the various arguments pro and con are beyond the scope of this book, one recent development in particular warrants consideration here.

In the 2010 edition of its annual *World Energy Outlook* (released in November 2010), IEA—which has generally been quite optimistic about future energy supply—explicitly discussed the issue of peak oil for the first time. Partly on the basis of its own field-by-field survey of oil reserves in 2008, IEA estimated that worldwide production of crude oil from *conventional* sources most likely peaked in 2006.[11]

IEA expects that unconventional oil (from tar sands, the Arctic, and deepwater fields) plus NGLs can offset declining conventional oil and enable increased world oil production for two more decades. However, the agency warns that significant investment will be required to achieve this goal.

Global oil consumption dropped from 86.2 million barrels per day in 2007 to 84.2 million in 2009, reflecting the global economic downturn. However, a projection by OPEC in July 2011 estimated that global demand would increase by 1.36 million barrels per day in 2011 over 2010, to an expected total of 88.2 million barrels for the full year, a new record high.[10] Looking to 2012, IEA has projected another record high consumption level of 91.0 million barrels.[9]

Final Thoughts

Specific details and a timetable for the evolution of the petroleum industry are impossible to predict. Nevertheless, it can be said that the early 21st century is already creating the conditions—as well as the requirement—for innovation.

New thinking, new practices, and new technology will be needed to address the growing confluence of concern about climate change, energy security, resource nationalization, and rising energy prices. Success will come to those in the industry who put innovation at the core of their business strategies, upstream, midstream or downstream—from the wellhead to the point of use of oil, natural gas, and their many derivative products.

References

1 U.S. Energy Information Administration. *International Energy Outlook 2011.* http://www.eia.gov/oiaf/ieo/world.html (accessed September 30, 2011).

2 Ibid.

3 Ibid.

4 Sloan, L. 2010. The challenges facing America's 21st century oil and gas workforce. In *2010 International Petroleum Encyclopedia.* Tulsa, OK: PennWell.

5 Saunders, B. The oil & gas industry's great crew change (series): What's expected to happen, and how companies are preparing. September 2011. *Rigzone,* Houston, TX.

6 Houlton, G. (Purvin & Gertz; Houston). 2010. Crude demand to increase, feed-quality changes in store. In *Oil & Gas Journal.* December 6: 119. Tulsa, OK: PennWell.

7 Bowden, A. The economic cost of maritime piracy: Working paper. December 2010. Broomfield, CO: One Earth Future Foundation.

8 BP. 2011. *Energy Outlook 2030.* http:www.bp.com. BP. London

9 IEA. 2010. *World Energy Outlook 2010.* Paris.

10 OPEC. Monthly oil market report. July 2011. Vienna.

11 IEA. 2010.

Terms, Abbreviations, and Acronyms

2D, 3D, 4D: two-, three-, four-dimensional seismic

AC: alternating current

AUV: autonomous underwater vehicle

AVA: amplitude variation with angle

AVO: amplitude variation with offset

b: billion

BHA: bottom-hole assembly

bbl: barrel

bcf: billion cubic feet

bcm: billion cubic meters

boe: barrels of oil equivalent

BOP: blowout preventer

bpd (b/d): barrels per day

bph: barrels per hour

BS&W: basic sediment and water

Btu: British thermal unit

capex: capital expenditure

cd: calendar day

cf: cubic feet

CFC: chlorofluorocarbon

CO_2: carbon dioxide

CSEM: controlled-source electromagnetic

CT: compliant tower; *or,* computerized tomography

CTD: coiled-tubing drilling

CTE: coal tar enamel

DSU: drilling spacing unit

ERW: electric resistance welding

DA: direct assessment

DC: direct current

DP: dynamic positioning

DST: drill stem test

DSV: diver support vessel

EM: electromagnetic

EOR: enhanced oil recovery

ERD: extended-reach drilling

ESP: electrical submersible pump

FBE: fusion-bond epoxy

FDPSO: floating, drilling, production, storage, and off-loading vessel

FEED: front-end engineering and design

FEWD: formation evaluation while drilling

FMEA: failure modes and effects analysis

FPS: floating production system

FPSO: floating production, storage, and off-loading vessel

FSO: floating storage and off-loading vessel

ft: feet

F-T: Fischer-Tropsch

gal: gallon

GLR: gas : liquid ratio

GOC: gas-oil contact

GOM: Gulf of Mexico

GOR: gas : oil ratio

gph: gallons per hour

GPS: Global Positioning System

GTL: gas to liquids

HCWC: hydrocarbon–water contact

HDD: horizontal directional drilling

HPHT: high-pressure/high-temperature

H$_2$S: hydrogen sulfide

HSE: health, safety, and the environment

HSP: hydraulic submersible pump

Hz: hertz (cycles per second)

ILI: internal line inspection

IMP: integrity management plan

in.: inches

IOC: international oil company

km: kilometers

LDC: local distribution company

LNG: liquefied natural gas

LPG: liquefied petroleum gas

LWD: logging while drilling

m: meters

M: thousand

MAOP: maximum allowable operating pressure

Mcf: thousand cubic feet

Mcfd: thousand cubic feet per day

MDT: thousand decatherms

MFL: magnetic flux leakage

mi.: miles

MM: million

MODU: mobile offshore drilling unit

MSV: multipurpose service vessel

MT: metric ton

MWD: measurement while drilling

NGL: natural gas liquids

NGO: nongovernmental organization

NMR: nuclear magnetic resonance

NOC: national oil company

OBM: oil-based mud

OD: outside diameter

OWC: oil-water contact

PD: positive displacement

PDC: polycrystalline diamond compact

PLT: production logging tool

ppm: parts per million

PSA: production-sharing agreement

PSC: production-sharing contract

psi: pounds per square inch

psia: pounds per square inch absolute

psig: pounds per square inch gauge

PVT: pressure volume and temperature

RF: recovery factor

ROP: rate of penetration

ROV: remotely operated vehicle

ROW: right-of-way

rpm: revolutions per minute

RVP: Reid vapor pressure; *or,* pounds Reid vapor pressure

SBHP: static bottom-hole pressure

SBM: single-buoy mooring

SCADA: supervisory control and data acquisition

SCC: stress corrosion cracking

scf: standard cubic foot

SI: International System of Units

SiO$_2$: silicon dioxide (quartz)

SMYS: specified minimum yield strength

SNG: synthetic natural gas

SOBM: synthetic oil-based mud

SPR: strategic petroleum reserve

STP: standard conditions of temperature and pressure

t: tons

TAPS: Trans-Alaska Pipeline System

tcf: trillion cubic feet

TD: total depth

TEG: triethylene glycol

TLP: tension-leg platform

ULCC: ultralarge crude carrier

UR: ultimate recovery

USD: U.S. dollars

UT: ultrasonic testing

UV: ultraviolet

VLCC: very large crude carrier

VSP: vertical seismic profiling

WBM: water-based mud

WOB: weight on bit

WOC: wait on cement

WTI: West Texas Intermediate

Further Reading

Technical Books and Professional Articles

Ahlbrandt, T. S., R. R. Charpentier, T. R. Klett, J. W. Schmoker, C. J. Schenk, and G. F. Ulmishek. 2005. *Global Resource Estimates from Total Petroleum Systems*. Alexandria, VA: American Association of Petroleum Geologists.

Ahmed, T. 2006. *Reservoir Engineering Handbook*. 3rd ed. Oxford, UK: Gulf.

American Petroleum Institute. 2008. Specification for pipeline. API 5L. 44th ed. Washington, DC: American Petroleum Institute.

Archer, J. S., and C. G. Wall. 1999. *Petroleum Engineering: Principles and Practice*. Berlin: Kluwer Academic.

Azar, J. J., and G. R. Samuel. 2007. *Drilling Engineering*. Tulsa, OK: PennWell.

Brantly, J. E. 1971. *History of Oil Well Drilling*. Houston: Gulf Publishing Co.

Burdick, D. L., and W. L. Leffler. 2001. *Petrochemicals in Nontechnical Language*. 3rd ed. Tulsa, OK: PennWell.

Busby, R. L. 1999. *Natural Gas in Nontechnical Language*. Tulsa, OK: PennWell.

Center for Energy Economics at the Bureau of Economic Geology. 2007. *Introduction to LNG*. Austin, TX: University of Texas at Austin.

Consentino, L. 2001. *Integrated Reservoir Studies*. Paris: Institut Français du Pétrole.

Dake, L. P. 2004. *The Practice of Reservoir Engineering*. Revised ed. Oxford, UK: Elsevier.

Devereux, S. 1999. *Drilling Technology in Nontechnical Language.* Tulsa, OK: PennWell.

Economides, M., L. Watters, and S. Dunn-Norman. 1997. *Petroleum Well Construction.* Hoboken, NJ: Wiley.

Gluyas, J., and R. Swarbrick. 2003. *Petroleum Geoscience.* Oxford, UK: Blackwell Science.

Hyne, N. J. 2001. *Nontechnical Guide to Petroleum Geology, Exploration, Drilling, and Production.* 2nd ed. Tulsa, OK: PennWell.

Jahn, F., M. Cook, and M. Graham. 2008. *Hydrocarbon Exploration and Production.* 2nd ed. Oxford, UK: Elsevier.

Johnston, D. 1994. *International Petroleum Fiscal Systems and Production Sharing Contracts.* Tulsa, OK: PennWell.

Joshi, S. A. 1991. *Horizontal Well Technology.* Tulsa, OK: PennWell.

Kearey, P., M. Brooks, and L. Hill. 2002. *An Introduction to Geophysical Exploration.* 3rd ed. Oxford, UK: Blackwell Science.

Leffler, W. L. 2000. *Petroleum Refining in Nontechnical Language.* 3rd ed. Tulsa, OK: PennWell.

Leffler, W. L., R. A. Pattarozzi, and G. Sterling. 2003. *Deepwater Petroleum Exploration and Production: A Nontechnical Guide.* Tulsa, OK: PennWell.

Magoon, L. B., and W. G. Dow. 1994. *The Petroleum System: From Source to Trap.* Alexandria, VA: American Association of Petroleum Geologists.

McAleese, S. 2000. *Operational Aspects of Oil and Gas Well Testing.* Oxford, UK: Elsevier.

Miesner, T. O., and W. L. Leffler. 2006. *Oil & Gas Pipelines in Nontechnical Language.* Tulsa, OK: PennWell.

National Energy Board. 2006. *Canada's Oil Sands: Opportunities and Challenges to 2015.* Ottawa, ON: National Energy Board.

North, F. K. 1985. *Petroleum Geology.* London: Allen & Unwin.

Odell, P. R. 2004. *Why Carbon Fuels Will Dominate the 21st Century's Global Energy Economy.* Brentwood, UK: Multi-Science.

Parkash, S. 2003. *Refining Processes Handbook.* Oxford, UK: Gulf Professional Publishing (imprint of Elsevier, Inc.).

Perrin, D. 1999. *Well Completion and Servicing.* Paris: Institut Français du Pétrole.

Raymond, M. S., and W. L. Leffler. 2005. *Oil & Gas Production in Nontechnical Language.* Tulsa, OK: PennWell.

Rider, M. L. 1996. *The Geological Interpretation of Well Logs.* Caithness, Scotland: Whittles.

Selley, R. C. 1997. *Elements of Petroleum Geology.* 2nd ed. Tulsa, OK: PennWell.

Tiab, D., and E. Donaldson. 2004. *Petrophysics: Theory and Practice of Measuring Reservoir Rock and Fluid Transport Properties.* 2nd ed. Houston: Gulf Professional Publishing (imprint of Elsevier, Inc.)

Tiratsoo, E. N. 1986. *Oilfields of the World.* 3rd ed. Houston: Gulf Publishing Co.

Tusiani, M. D. 1996. *The Petroleum Shipping Industry: Operations and Practices.* Tulsa, OK: PennWell.

Tusiani, M. D., and G. Shearer. 2007. *LNG: A Nontechnical Guide.* Tulsa, OK: PennWell.

General Interest

Bernstein, P. L. 1998. *Against the Gods: The Remarkable Story of Risk.* Hoboken, NJ: Wiley.

Campbell, C. J. 2005. *Oil Crisis.* Brentwood, UK: Multi-Science Publishing.

Cordesman, A. H., and K. R. al-Rodhan. 2006. *The Global Oil Market: Risks and Uncertainties.* Washington, DC: Center for Strategic and International Studies Press.

Deffeyes, K. S. 2001. *Hubbert's Peak: The Impending World Oil Shortage.* Princeton, NJ: Princeton University Press.

Downey, M. 2009. *Oil 101.* New York: Wooden Table.

Marcel, V. 2006. *Oil Titans: National Oil Companies in the Middle East.* London: Chatham House.

Margonelli, L. 2007. *Oil on the Brain: Adventures from the Pump to the Pipeline.* New York: Nan A. Talese/Doubleday.

Maugeri, L. 2006. *The Age of Oil: The Mythology, History, and Future of the World's Most Controversial Resource.* Westport, CT: Praeger.

Simmons, M. R. 2005. *Twilight in the Desert: The Coming Saudi Oil Shock and the World Economy.* Hoboken, NJ: Wiley.

Smil, V. 2008. *Oil.* Oxford, UK: Oneworld.

Van Vactor, S. 2010. *Introduction to the Global Oil & Gas Business.* Tulsa, OK: PennWell.

Yergin, D. 1991. *The Prize: The Epic Quest for Oil, Money and Power.* New York: Simon & Schuster.

Yergin, D. 2011. *The Quest: Energy, Security and the Remaking of the Modern World.* New York: Simon & Schuster.

Other Information Sources

Baker Hughes. http://www.bakerhughes.com/

BP *Statistical Review of World Energy.* http://www.bp.com/

ODS-Petrodata. http://www.ods-petrodata.com/

Oil & Gas Journal. http://www.ogj.com/

Rigzone. http://www.rigzone.com/

World Oil. http://www.worldoil.com/

Organizations, Agencies, and Professional Societies

Trade Associations and Professional Societies

American Association of Petroleum Geologists (AAPG),
http://www.aapg.org/

American Gas Association (AGA), http://www.aga.org/

American Petroleum Institute (API), http://www.api.org/

American Society of Mechanical Engineers (ASME)http://www.asme.org/

Association of Oil Pipe Lines (AOPL), http://www.aopl.org/

ASTM International (formerly American Society for Testing and
Materials), http://www.astm.org/

Gas Processors Association (GPA), http://www.gasprocessors.com/

Gas Technology Institute (GTI) (formerly Gas Research Institute [GRI]),
http://www.gastechnology.org/

Independent Petroleum Association of America (IPAA),
http://www.ipaa.org/

Interstate Natural Gas Association of American (INGAA),
http://www.ingaa.org/

Interstate Oil and Gas Compact Commission (IOGCC),
http://www.iogcc.state.ok.us/

Natural Gas Supply Association (NGSA), http://www.ngsa.org/

SAE International (formerly Society of Automotive Engineers),
http://www.sae.org/

Society of Economic Geologists, http://www.segweb.org/

Society of Exploration Geophysicists, http://www.seg.org/

Society of Petroleum Engineers (SPE), http://www.spe.org/

U.S. Government Agencies

Bureau of Labor Statistics (BLS) (part of the Department of Labor), http://www.bls.gov/

Bureau of Ocean Energy Management, Regulation and Enforcement (BOEMRE) (formerly Minerals Management Service; part of the Department of the Interior), http://www.boemre.gov/

Department of Energy (DOE), http://www.energy.gov/

Department of the Interior (DOI), http://www.doi.gov/

Department of Transportation (DOT), http://www.dot.gov/

Energy Information Administration (EIA) (part of the Department of Energy), http://www.eia.doe.gov/

Environmental Protection Agency (EPA), http://www.epa.gov/

National Oceanic and Atmospheric Administration (NOAA) (part of the Department of Commerce), http://www.nws.noaa.gov/

National Transportation Safety Board (NTSB), http://www.ntsb.gov/

Occupational Safety and Health Administration (OSHA) (part of the Department of Labor), http://www.osha.gov/

Pipeline and Hazardous Materials Safety Administration (PHMSA) (part of the Department of Transportation), http://www.phmsa.dot.gov/

International Energy Organizations

International Energy Agency (IEA) (part of the Organization for Economic Cooperation and Development [OECD]), http://www.iea.org/

International Maritime Organization (IMO) (part of the United Nations), http://www.imo.org/

International Organization for Standardization (ISO), http://www.iso.org/

National Energy Board (NEB) (Canada), http://www.neb-one.gc.ca/

Organization of the Petroleum Exporting Countries (OPEC), http://www.opec.org/

Index